AF251538

FIRE
INSTRUCTOR'S
TRAINING GUIDE

About the Author

Joe Bachtler began his lifetime affiliation with firefighting as a volunteer in suburban Philadelphia in 1943. Following service as a firefighter with the United States Marine Corps, he completed the Fire Technology program at Oklahoma State University (then Oklahoma A. & M. College) and earned his B.S. degree in public administration, fire administration major, from the University of Southern California in 1956. During this period he also worked as a firefighter with the city of Stillwater, Oklahoma, and later was appointed to the Burbank, California, Fire Department, where he was assigned to truck 1, Headquarters Station.

In the late 1950s and early 1960s he served as fire chief for Callery Chemical Company, Pittsburgh, Pennsylvania, and as fire chief at the Gary Steel Works, United States Steel Corporation, in Gary, Indiana.

In 1962 Bachtler joined the staff of the Fire Service Extension (now Maryland Fire and Rescue Institute), University of Maryland. For the past twenty-five years he has been actively involved and nationally recognized as a teacher, speaker, and writer in fire service education and training. In 1975, after serving as director of the Maryland State Fire Training Program, he joined the staff of the National Fire Academy, where he served as education and training supervisor and later as associate superintendent. He returned to the Maryland Fire and Rescue Institute in 1979 and is currently involved with program development in the Institute Development Division. As an instructor-trainer since 1966, Bachtler has taught scores of fire, rescue, and emergency-care instructors in his home state of Maryland and in many other parts of the nation.

Married thirty-five years, Bachtler and his wife, Dede, have three daughters and two grandchildren. They currently live, and plan soon to retire, on Maryland's Eastern Shore in the city of Salisbury. Dede is a freelance artist, best known to people in the fire service for the twenty-five years of elegant Christmas covers she created for the Maryland Fire and Rescue *Bulletin* from 1962 to 1986.

A licensed racing driver, Bachtler maintains and drives a Formula Vee in Washington area Sports Car Club of America road racing events. He is lay leader in Salisbury's Grace United Methodist Church, where he also sings in the choir and teaches an adult Sunday school class. He is a frequent guest speaker at Maryland prayer breakfasts and in Wicomico County churches.

FIRE INSTRUCTOR'S TRAINING GUIDE

SECOND EDITION

J.R. Bachtler

A PennWell Publication

Library of Congress Cataloging-in-Publication Data

Bachtler, J.R., 1928–
    Fire instructor's training guide.

    "A PennWell publication."
    Includes index.
    1. Fire prevention—Study and teaching.   I. Title.
TH9120.B33   1988   628.9′2′071   88-63259
ISBN 0-87814-912-0

THIS BOOK IS DEDICATED
TO MY 2400 M STREET COLLEAGUES,
WITH WHOM I SHARED THOSE EARLY, CHAOTIC DAYS
IN WHICH
THE NATIONAL FIRE ACADEMY WAS CREATED:
IN PARTICULAR,
TO THAT SMALL AND ELITE GROUP
OF FIRE SERVICE EDUCATORS
WHO GAVE SO GREATLY OF THEMSELVES
AND THEN MOVED ON,
WITHOUT RECOGNITION OR THANKS
FOR THEIR IMMENSE CONTRIBUTIONS

Contents

Foreword

The *Fire Instructor's Training Guide* was written to fill a perceived need in the early 1970s. That need focused on the development of instructional skills necessary for the burgeoning number of firefighters and fire officers filling the fire department instructor ranks in fire departments throughout the country. Professionalism and skill proficiency were on the rise, as were instructor interests and capabilities.

I'd like to think that the first edition served well and filled a need, at least in part.

In the late sixties and early seventies, community colleges and two-year fire programs were springing up in virtually every state, the National Fire Prevention and Control Commission was developing a national fire focus, the creation of the Joint Council of National Fire Service Organizations was close at hand, and, in the not too distant future, the establishment of the National Fire Academy, professional qualifications, and various levels of fire service certification awaited. OSHA, NIOSH, SARA, and NFPA 1500 would appear much later on the fire service scene.

I recall these developments because, individually, each has a profound effect on the fire service and, collectively, on what we are mandated to do. How we go about doing it and with whom we have to work have changed dramatically.

Why is the second edition of *Fire Instructor's Training Guide* necessary at this time? The French expression, "The more things change, the more they remain the same," could not be more appropriate.

New training skills, techniques, and strategies applied to tested and experienced methodology will, it is hoped, serve a new breed of enthusiastic fire service instructor and training officer. And what better time than now for the author of the first edition to pass on the instructional challenge to a more contemporary instructor's instructor?

Joe Bachtler's second edition of the *Fire Instructor's Training Guide* is the right author and the right text for the instructional requirements of this year and next. Joe Bachtler, through his own fire service experience and continuous involvement in instructor development, has provided one of the best instructor references ever written. The reader is patiently walked through objective and goal setting, instructional methodology, lesson plan development, and instructional computer technology—to name but a few—by a master instructor whose experience and common sense savvy breathe fresh vitality into a tired subject.

My hat's off to Joe Bachtler and, more important, to the reader who is charged with the responsibility of taking the fire service to the next plateau. Leave the fire service better than you found it . . . it's your responsibility and duty!

Good luck!

Anthony R. Granito
Superintendent
Nassau County Fire Service Academy

Preface

As Chick Granito indicated in his gracious foreword, this second edition of *Fire Instructor's Training Guide* is substantially different from the first. The fire service has indeed changed dramatically over the past two decades. Concurrently, the status and the needs of fire service instructors have changed along with our profession. If there ever was any question of the importance of the role played by the fire service instructor, that question has been put to rest by the establishment of the National Professional Qualifications System, which recognizes teaching as a separate and distinct area of fire service competence with highly specific standards covering a special body of knowledge.

While I made no attempt to write to the *Standard for Fire Service Instructor Professional Qualifications*, NFPA 1041, the reader will note that, with very minor differences in terminology, this book addresses all of Instructor I and Instructor II, most of Instructor III, and the instructional aspects of Instructor IV. It is, then, compatible with the standard and can serve as a valuable resource for those interested in meeting the requirements which 1041 establishes.

In addition to this fundamental information, I have included extended discussions of several topics beyond the level found in most instructor training programs and publications. Course development in the political arena, lesson plan format and design, levels of instruction, the writing of proper performance objectives, the specifics of projected simulations, and special application techniques are all presented in substantial "how to" detail. In our experience, each of these areas presents particular problems for many fire service instructional agencies and therefore merits special attention.

In combining very basic instructor training information with a number of relatively advanced concepts and procedures, our objective has been to create a technical resource which will meet a wide spectrum of instructional needs. If you find yourself reaching for this book from time to time as instructional problems arise in the years ahead, that goal will have been well met.

My only certain claim to originality is in the arrangement and presentation of the material which appears herein. One of the most important factors in the growth and strength of fire service training in this nation has been the complete openness with which instructors and instructor trainers have shared their ideas. During the third of a century since I took my first instructor training course, information has been freely exchange, modified, and exchanged again, until it is virtually impossible to say with any degree of confidence who originated what. Among others, I have been greatly influenced by the late Professor R.J. Douglas, Fire Protection Technology, Oklahoma A&M College; the late George L. Orgain, former instructor trainer, Oklahoma State Fire Service Training; Robert C. Byrus, former director, Fire Service Extension, University of Maryland, now retired; Sherman A. Pickard, former director, Fire and Rescue Service Training, North Carolina Department of Insurance, now fire chief, Raleigh, N.C.; William L. Brannan, former instructor trainer, United States Internal Revenue Service, now retired; and Jesse V. Jackson, associate director, and Thomas C. Cusick, instructor trainer, both currently with the Maryland Fire and Rescue Institute.

I owe a special debt to John W. Hoglund, director, Maryland Fire and Rescue Institute, for the standards of excellence he established in the 1960s as a dedicated trainer of instructors, and for the knowledge and understanding of teaching and learning which he passed on to those of us who followed him in the Maryland Instructor Training Program. I especially thank him for his support and his encouragement during the writing of this book.

My sincere gratitude is given to Thomas W. Wilson, Field Operations Division, Maryland Fire and Rescue Institute, for his painstaking review of the final manuscript and for his assistance in helping me ensure that my words actually say what I intended to say. Thanks are also extended to John C. Hess, Maryland Fire and Rescue Institute Administration, for his advice in the area of computers and his review of the computer chapter. To our editor, Rose Jacobowitz, I extend my deep admiration for her wisdom, her attention to detail, and for her keen interest and generous assistance. Any errors in grammar and style are mine, not hers. And let's not forget Art Arias and his great drawings that add so much to the text.

Finally, my special thanks go to Anthony R. "Chick" Granito, for inviting me to write a sequel to his first edition of *Fire Instructor's Training Guide*, and for his enthusiasm, advice, and counsel during the preparation of this manuscript.

J.R. BACHTLER

Salisbury, Maryland
July 1988

1

Fire Instructors and Fire Service Students

Come in! Sit down. We've been looking forward to talking with you. We understand you're interested in fire service instruction and we think that's just great. We've been kicking around in this business for a long time, and we're delighted to have this chance to share some of the ideas we've picked up along the way. We're convinced that instructors are some of the most important people in the fire service. Oh, sure, you can learn without an instructor; we've even seen cases where students had to learn *in spite* of the instructor, but that situation doesn't change our opinion. Whatever your rank, whatever your experience, whatever the size of your department, your role as a fire service instructor is far more important than you probably realize.

The Influence of the Fire Instructor

Fire instructors come in all sorts of ranks, sizes, shapes, and ages. Most of them share some extremely valuable traits. Most fire instructors are relatively experienced, most love the fire service, and most are vitally interested in doing a first-rate job for the students they train and for the department in which they serve. *All* of them share something else: an

Fig. 1-1. Teaching and administration—partners in progress.

exceptional influence within their respective municipal departments or counties or states—an influence which shapes the fire service of today and establishes the direction and the goals for the fire service of tomorrow.

The effect of an instructor has little or no relationship to the instructor's rank. The "Training Officer" of a large metropolitan department may be a very senior chief officer. In a small paid or volunteer department, the person responsible for training may be a captain or a lieutenant. The common denominator is that the person in charge of training, be it senior chief or one-trumpet lieutenant, has an influence within the department that is second only to that of the Fire Chief (Fig. 1-1). In different settings, the county and state fire service instructors share this same level of effect and responsibility.

Fire instructors in every rank of the command chain, from private on up, provide the policy makers with feedback, suggestions, and ideas on which many operational changes are based. The person in charge of training directly influences the content and the style of training and thereby, with or without intent, creates operational policy. Every member of the instructional staff will, in some fashion, affect the operations of the department or departments in which that member teaches through attitudes, the personal slant placed on the training material, and the kind of instruction he or she provides.

Influence as a Change Agent

Fire instructors at every level of rank and experience share in this unique opportunity to act as "change agents." The instructor has the first contact with new recruits: people new to the fire service, people at an *impressionable age,* and people at the most *impressionable stage* of their careers. New firefighters look for role models, and the role models their instructors provide will live on, long after the instructors themselves retire. The legacy of the impressions instructors give to new personnel may be constructive or it may be destructive. In either case, the recruits these instructors train will influence the department, for good or for ill, during many years to come.

Incidentally, one of the true challenges of instruction is maintaining the ability to relate effectively to brand-new, totally green personnel. The longer each of us is in the fire service, the more difficult it becomes to remember just how confusing and overwhelming our profession is to the novice. There will *always* be the need for one of us to stand by a ladder, in the center of a circle of youngsters, and say, "This is the beam and these are the rungs; that part up there is the tip, and this is the heel." Try not to lose the ability to communicate effectively at the most basic level. One of the most rewarding aspects of teaching is to stand back at the end of a recruit class and watch your students climb with confidence or swing into a piece of breathing apparatus as if it were an old friend. It's a good feeling.

Instructors continue to exert influence at every career stage through which firefighters pass. The relationships change, of course, and, in many respects, our teaching job becomes more difficult when we are instructing experienced personnel in advanced subjects, but the rewards grow with the challenge. Providing in-service or advanced training gives us an opportunity to help people grow in the job, to help them become mature and effective firefighters and fire officers. When we assist them in preparing for advancement, we can have no idea what the eventual benefit to the department will be. We have been teaching long enough to be able to state with confidence that the true reward of being a fire service instructor comes late in your teaching career, when you can look at a number of the fire service leaders in your department or county or state and realize that you contributed, in some small measure, to their success. That's a good feeling, too.

Influence on Safety

One final thought about the influence of the fire service instructor concerns safety. We'll not dwell on it here because we will discuss it in depth later, but be aware of the importance of your role. *Your* attitude toward safety and the way you conduct *yourself* on the drill ground and at emergency scenes will do a great deal to establish the total attitude of your people toward safety. If you honestly want to help them to live, uninjured, as long as possible, *you* must set the example, both in your teaching and in your personal actions. Bob Byrus, former Director of the Maryland Fire Service Extension, told his instructors on many occasions, "Don't let the ghost of an untrained fire-fighter come back to haunt you!"—not a bad piece of advice.

Our Work Is Difficult

Providing meaningful and effective fire service education and training is not an easy task. The work we teach is both broad in scope and, at the same time, extremely detailed. If you doubt that statement, stop for a moment the next time you pass an apparatus pump panel. Count the number of knobs and levers and gages, and consider everything the operator of that pump must know in order to utilize the equipment effectively; then think of what that pump operator had to learn *before* moving from the rank of back-step firefighter into the left-hand front seat of that piece of fire apparatus (Fig. 1-2).

In addition to being highly detailed, much of what we must convey to our students is, in all honesty, boring. Certainly a lot of fire service training is fascinating; working in breathing apparatus, working at extreme height, working in close proximity to fire are all exciting and challenging experiences, but that doesn't change the fact that a lot of the rest of what must be learned *is* boring. The nomenclature of ladders, folding salvage covers, rules and regulations, fixed fire protection systems, fire service organization and administration are not, in the minds of most firefighters, dynamic subjects, but they "go with the territory." Motivation, then, becomes an important part of our job, and motivation is a topic we will return to again and again. Motivation has been described as getting a person to do something and making the individual believe he or she *wants* to do it. If that kind of keen motivation is not present, it is our job to try to stimulate it in our students. Fortunately for us, this process isn't too difficult. Fire service people, as you well know, are usually highly motivated to begin with.

Finally, fire service training is frequently quite repetitive. We do the same things over and over and over again. We *must* require repetition to bring our beginners to an acceptable level of proficiency, and during in-service training we must require repetition to hone and maintain emergency skills, particularly those which are infrequently used. Like all human beings,

Fig. 1-2. Remember *your* rookie days?

firefighters resist doing the same task over and over again. "We've already done that!" "We did that last week (*or* last month, *or* last year, as the case may be)!" These are complaints you can expect to hear and yet, somehow, through motivation and explanation and patient encouragement, you must get the job done, done well, and done on time. Our work *is* difficult.

Our Work Is Different

Teaching in the fire service is different from the teaching we have experienced in secondary schools and colleges. In most schools, 70 percent is considered "passing." What is "passing," when you are teaching someone to operate that pumper we mentioned in the last section? What is "passing," when you are teaching breathing apparatus or, perhaps, CPR? In each of these cases, every single element is critical, and every single element must be learned perfectly. We may establish an inclusive passing grade for some segments of a training program, but specific tasks and functions within that program cannot be left at 70 percent proficiency, or someone is going to be injured or killed.

In conventional education, teachers and instructors often adopt a "take it or leave it" attitude. In many settings, particularly at the college level, there are instructors and department heads who do not care at all if an individual student succeeds or fails. We do not have that luxury. We fire service instructors must be absolutely dedicated to ensuring, to the best of *our* ability, that our students *do* achieve, to the best of *their* ability.

The level of responsibility we assume is rather frightening. Some years ago, Dr. John L. Bryan, professor and head of the Fire Protection Engineering Curriculum at the University of Maryland, addressed the Fire Department Instructors Conference at Memphis. In his presentation, Dr. Bryan stated to his audience of fire service instructors: "Your students must involuntarily recall, in times of great emotional and physical stress, that which they need to know." That statement is extremely wise, and it is packed full of ideas we should note and note well. "Your students," not ours or his or hers but *yours*, the people *you* train, the people *you* are responsible for, are the ones about whom Dr. Bryan is talking. Your students "must," not should, or ought to; *your* students *must*. Your students must "involuntarily recall"; they must recall without thinking, they must react automatically, they must perform their jobs completely and perfectly, as if by reflex action. "In times of great emotional and physical stress" is a phrase which should be understandable to any experienced firefighter (Fig. 1-3). It doesn't happen on every run, by any means, but the times of great emotional and physical stress are frequent enough so that we can instantly relate to the point Dr. Bryan was making. The people we teach must be able to recall involuntarily that which they need to know, regardless of the circumstances they encounter, and no person who has *not* been in our business could possibly comprehend just how stressful those circumstances can be. The final phrase, "that which they need to know," stands for an immense quantity of information. In just six words, Dr. Bryan has charged us, the fire service instructors, with responsibility for transmitting to our students the entire body of fire service knowledge they must have to do their jobs. We said earlier that the responsibility is frightening. Perhaps "awesome" would be a better word. Our job *is* difficult, and it *is* different.

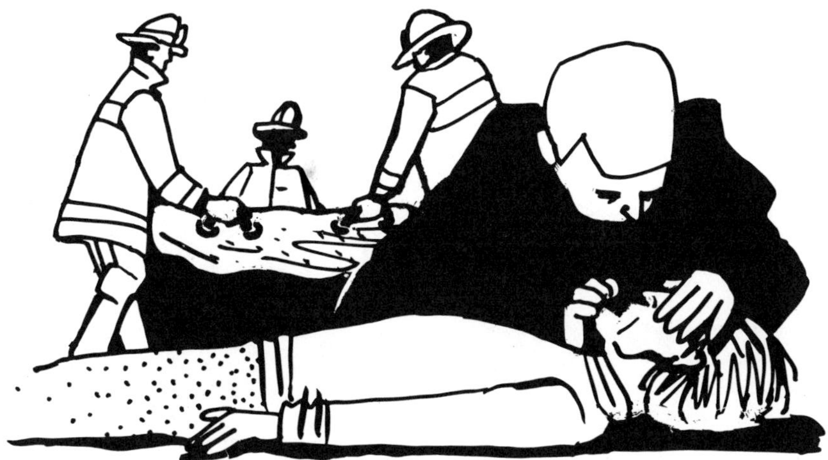

Fig. 1-3. "Times of great emotional and physical stress . . . "

Two Sets of Tools

To perform effectively as an instructor, it has been said that we need two sets of tools (Fig. 1-4). The first set contains the information necessary within our profession. We assume you already have that set of tools. No two of us are going to have exactly the same educational background nor the same quantity or quality of emergency experience, but each of us, if we are preparing to enter or are working in fire service education and training, most probably have professional credentials appropriate for the fire service community in which we serve.

The second set of tools contains the knowledge and techniques of teaching. That, we assume you do not have. This book is written on the assumption that you have been assigned or have inherited instructional responsibility with little or no opportunity for formal instructor training. If you have been fortunate enough to participate in a good, comprehensive instructor development program, some of the material in this book will be redundant. We hope, in either case, you will find what we have written to be both valid and useful.

Your Duty to Your Students

It has become almost traditional in instructor training to mention, in general terms, the instructor's duty to the students, to the administration,

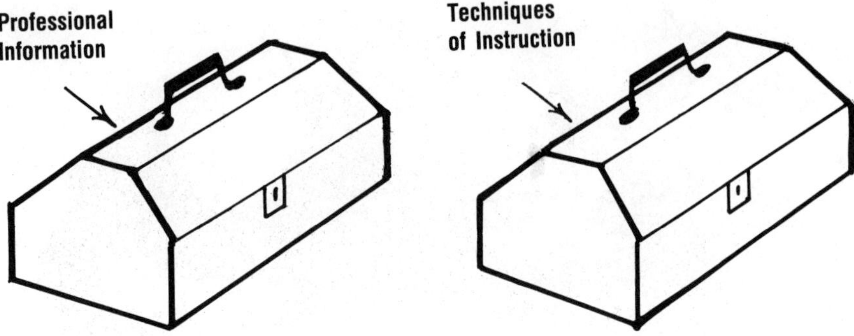

Fig. 1-4. Two sets of tools.

and to self. Your obligation to your students is to see that they *learn*. You can't force them to learn, of course, and you can't personally make up for any really serious deficiencies in student ability, but that does not relieve you of the obligation to provide a learning *experience* and a learning *environment* which give the serious and capable firefighter a clear, unmistakable opportunity to succeed.

Your duty to students includes coming to class thoroughly prepared and ready to teach (Fig. 1-5). Come to class rested and with a level of vitality and vigor that permits you to give of yourself. Get there early. Get your aids and your equipment set up and checked out in advance. If you drag yourself through the classroom door at the last minute, tired and unprepared, you are cheating the very people you are there to serve: your students. There is an old saying: "If the student has not learned, the teacher has not taught." As with most old sayings, there is more than a little truth in it. If your students aren't learning, the first place to examine "why" is on *your* side of the lectern.

You have a duty to your students to know what you are talking about, to know your subject thoroughly and precisely. You need to keep up to date, to follow the professional changes which are constantly occurring, and to revise your lessons accordingly. You can*not* teach to someone else something that you do not know yourself. Sherman Pickard, then director of the North Carolina Fire Training Program, once put it in country dialect when he said, "You cain't no more teach what you ain't learned than you can go back to where you ain't been." We couldn't say it better.

Your Duty to Your Administration

You have many responsibilities to your administration. If you are paid,

Fig. 1-5. Vital instructor versus dragging instructor.

you have a basic duty to provide an honest hour of work for each hour of pay. You have an obligation to use time and equipment and materials carefully and economically. If you spend one hour teaching twenty students, you and the class expend a total of twenty-one hours during that one sixty-minute segment. In a paid department, twenty-one hours times the hourly rate of pay of the personnel involved is a significant sum of money. If you are working with volunteers, in that sixty minutes you have absorbed twenty hours which your students are donating to the community. Use those hours profitably.

You have a responsibility to carry out the policies of your administration. You will often be called upon by students to interpret official policies, and you *must* support those policies no matter what personal feelings you may have. There is no option; you have no choice. In the classroom, you *are* the administration.

We are not suggesting that you cannot or should not have your own ideas or opinions. If you find something in the training program, or in the policies you must teach, that you strongly oppose, we urge you to go and discuss it with the appropriate authorities within your organization. Give them the benefit of your thinking; give them your perspective. Argue; if necessary, *fight* for what you believe is right, but *not* in front of your class. Thrash out your differences of opinion behind a closed door, away from the students (Fig. 1-6). When that door opens, regardless of the final decision, *you* come

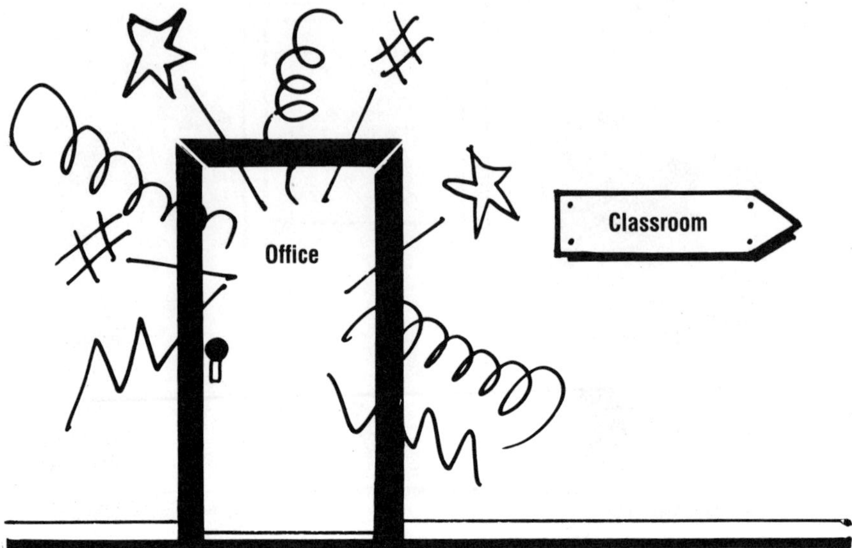

Fig. 1-6. Do your internal fighting away from the classroom.

out of the meeting and back to the classroom in total support of the established policy. If you find you cannot do so, we earnestly suggest you look for somewhere else to teach. An ounce of loyalty, it has been said, is worth a pound of cleverness.

You have a responsibility both to your students and to your administration to support actively all applicable equal opportunity laws and regulations. It is unlawful to discriminate against any person on the basis of race, color, creed, sex, marital status, personal appearance, age, national origin, or political affiliation. In testing and training, we must make sure that standards are applied objectively and consistently, and that all required standards are demonstrably job-related. In your professional and personal relationships with *all* your students, refrain from any form of ethnic or sexist remark. To do otherwise is offensive, unnecessary, and illegal.

Your Duty to Yourself

Finally, you have a duty to yourself. Teaching is a draining and difficult job. Do not spread yourself too thin. Part-time instructors in county and state training programs are particularly prone to scheduling class after class, night after night, until they are personally exhausted and their families are ready, indeed eager, to disown them. The better the instructor, the more it is likely

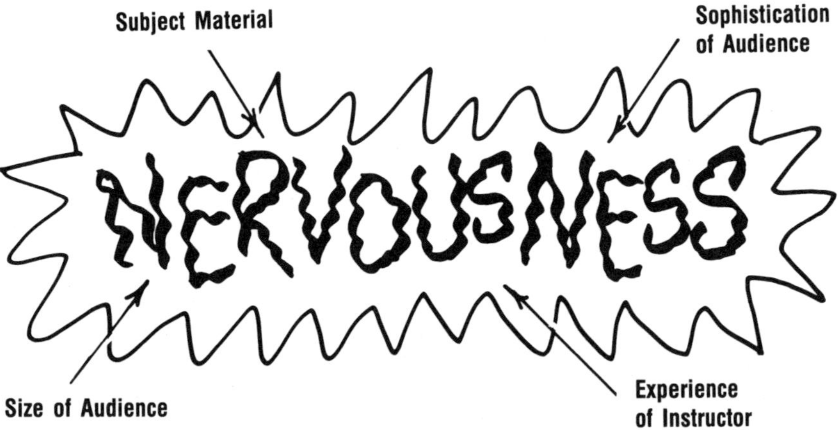

Fig. 1-7. Many things affect nervousness.

that he or she will be overworked. We have seen instructors attempt to carry on when they were overwhelmed with ill health or work or family or emotional problems. Don't let that happen to you. Take on as much as you can carry comfortably but be ready to say "No," clearly and firmly, when it is appropriate. One person cannot do it all. Don't even try.

Controlling Nervousness

Self-control, beginning with not letting yourself become too loaded with work, is an essential quality for every instructor. One of the first things most of us must learn to control is that familiar feeling of nervousness which wells up when we step in front of a group. Nervousness is perfectly natural, and every one of us encounters it many times during our teaching careers. The size of the audience, the sophistication of the audience, the subject material, and our personal level of teaching experience all have an effect on just how nervous we feel in any given set of circumstances (Fig. 1-7).

Being nervous is an indication that you are aware of your responsibility and that you are concerned with doing a good job. Both of these feelings are understandable, and both can actually help to make you a better instructor, in the long run, if you respond to them positively.

We don't have any magic method of making nervousness disappear, but we can suggest a few techniques that will help. First, be thoroughly prepared. Master your subject and master your lesson plan. Be sure you know how you are going to start off. In particular, master your opening statements; memorize them, if you feel it necessary. Once the class begins, once the kick-off is behind you, your nervousness will usually fall away and leave you free to concentrate on making things happen. When you step through the classroom door, let there be no doubt in your mind that you have mastered the subject, that you are in control. Easier said than done, of course, but it *will* help, and it *does* become easier as you gain experience. Make sure your equipment, supplies, and audiovisuals are all in place, checked out, and ready to go. This precaution means you must get to the teaching site well ahead of time. There is no better way to guarantee you *will* be nervous than to arrive late, unprepared, or disorganized. (If you arrive late *and* unprepared *and* disorganized, please don't tell the class you read our book.)

When you become nervous, your physiological actions and reactions will speed up. Pulse and breathing become quicker, and you will usually talk faster. The voice tends to rise in pitch. If you sense that any or all of these processes are happening, take a deep breath and consciously slow down. Slow your pace of speaking; control your voice. It is almost always possible to regain control by being consciously deliberate. Don't hesitate to do so.

Finally, think seriously about your own attitude toward the class. At least part of our nervousness comes from a fear of what the students will think of us and our instruction. In general, fire service students are more interested in the subject than they are in the instructor. Our students seldom come to class with specific intent to put any of us "on trial." Remember, too, that your students have no idea of what you have planned; they have no idea of what is supposed to happen next. As in a stage show, the audience, your class, does not know the next line. If it does not come out exactly as you intended, nobody but you knows the difference and, harsh as it may sound, nobody but you probably cares. Don't be too hard on yourself.

Patience, Patience, Patience

Another aspect of self-control is the ability to remain calm when things are going wrong. Students will make mistakes. Students will "rub you the wrong way." Outsiders, people who have no earthly business there in the first place, will disrupt your class from time to time. Your normal and natural inclination will be to flare up. Don't. Patience is an absolute necessity.

Woven throughout fire service education and training is the thread of discipline—not strict military discipline, even though we are a paramilitary service, but the kind of discipline which comes from deep within the individual. We teach the need to follow orders promptly and without question. We teach the need to work together as a team, each individual subordinating self to the unit. We teach the absolute necessity of staying, of hanging in there when things get tough, so we can protect one another.

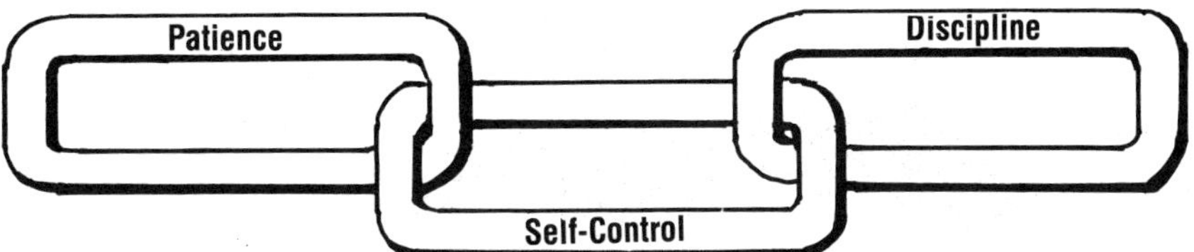

Fig. 1-8. Self-control, patience, and discipline are all linked together.

How can you model for your students this kind of internal self-discipline if *you* cannot discipline *yourself?* We don't believe you can (Fig. 1-8). Override your inclinations to explode. Treat the people around you with patience and tact and courtesy. It won't make you appear weak; it will convey a sense of strength. In particular, resist the temptation to return student stupidity or ineptitude with sarcasm or ridicule. It is easy to stand at the front of the room and make a student look foolish. The deck is stacked in your favor, and all the power is in your hands. Use it wisely, and use it gently. The person who is truly strong almost always exercises that strength carefully and quietly. The shouters and screamers (and we've all seen them at work in the fire service) are usually all noise and no thought, like a man Mark Twain once described. He said that this particular jackass reminded him of a little steamboat he once saw which had a five-foot boiler and an eight-foot whistle. "Whenever the whistle blew, the engine stopped" (Fig. 1-9). Keep your cool.

Controlling the Class

There are a lot of techniques by which a fire service instructor can control a class without making a lot of noise. You are the *de jure* leader by virtue of your assignment to that group of people as their instructor. The *de jure* leader leads by right, by the sanction of management, by regulation, by law. You have *that* asset right from the start. You can become the *de facto* leader, the leader in fact, by the way you fill the role.

Class Control Through Example

Look the part. Be concerned about your appearance, especially in the classroom setting. Your uniform should be clean and neat, complete as required by regulations, shoes shined, and your personal grooming correct (Fig. 1-10). On the drill ground, your turnout gear should be clean and in good repair. We'll talk a little more about that, later.

Fig. 1-9. Mark Twain's steamboat.

Let your enthusiasm show. We believe an instructor cannot have too much of it. How can you expect the class to be interested and enthusiastic about a subject if you are not? The same is true of pride; let that show too—not an overbearing personal pride, but pride in the fire service and pride in your organization. Let your students know they are part of a very special profession and that you, personally, are proud of the badge you wear. Let them know they are welcome to share in that pride, but *only* if they are willing to pay their dues along with the rest of us: dues that are payable in obedience and in service and in sweat and, from time to time, in blood and tears. We have a good "product." Use the positive aspects of our work to help keep your classes shaped up. Let this be absolutely clear. If they want to play in our game, they *will* play by our rules.

Class Control Through Session Discipline

Establish a sense of purposeful discipline in the classroom by starting on time (Fig. 1-11). If a class is set for 9:00 A.M. or 7:00 P.M., *start* at 9:00 A.M. or 7:00 P.M. *sharp*. If you wait for stragglers, people *will* straggle, and you will find yourself limping and stumbling into your teaching instead of starting with a bang. Students will quickly get the message and, after a session or two, we bet they will be in their seats and ready to go when the hour strikes. The other side of the coin is to let your people leave promptly. If clean-up

Fig. 1-10. Look the part.

runs ten minutes over one night, turn them loose ten minutes early at the end of the next session. Starting promptly and dismissing promptly set a businesslike tone for everything you do during the entire session.

Class Control Through Honesty

One sure way to *lose* control of a class is to fail to be truthful. Be honest with your students at all times. If you don't know some fact, say so; don't try to bluff your way through. Sooner or later you will be caught. There is no sin in saying, "I don't know, but I'll find out." Be truthful with your records and be honest in your responsibilities for quizzes and examinations. You must at all times require strict honesty on the part of your students; to do so you must be scrupulously honest yourself. Truth will never get you into trouble; honesty will never fail to pay off in the long run.

Class Control Through Teaching Techniques

You will communicate with your students primarily by your voice and your eyes. Using voice and eyes, our goal is to build a strong communication

Fig. 1-11. Start on time.

bridge between us and each student in the class. This communication link is absolutely essential if we are to be successful in transferring information. It is also extremely fragile, and it can be destroyed much more easily than it is built. A careless word, an ethnic or sexist remark, an antagonistic or sarcastic reply to a question, an implication that a student is unskilled or unintelligent—each can cause a barrier to fall between you and an individual student, or between you and the entire class. We are not suggesting that it is necessary to coddle students or to be hypersensitive about their feelings. We *are* suggesting that you be watchful and guard against words or actions which may make communication difficult. If you cannot communicate, you cannot teach. It's that simple.

Most of us, when we first attempt to teach, have a tendency to be overly formal. We can surely tell our students the words we want to say in this highly formal fashion, and our students will certainly hear them, but telling is not necessarily *teaching,* and listening is not necessarily *learning.* That is a *very* important distinction and we will come back to it in a slightly different context later on.

Teach Individuals, *Not* Classes

To establish personal contact, we must talk to individual students, not "the class." Stop and think for a moment. Recruit Class Number 35, or the Westend Pumps Class, or the Officer Candidate Class are just paper lists of names, existing only for administrative purposes (Fig. 1-12). Each of those "classes" is composed of a number of special, unique individuals, each with personal motivations, personal needs, and personal goals. Our job is to forge a direct link with each of them *as* individuals.

The best description we have heard is that you should *teach with all the earnestness of a private conversation.* Do your best to speak with the group in exactly the same relaxed manner, using the same voice inflections, the same simple words that you would use if you were talking with *one* of your students

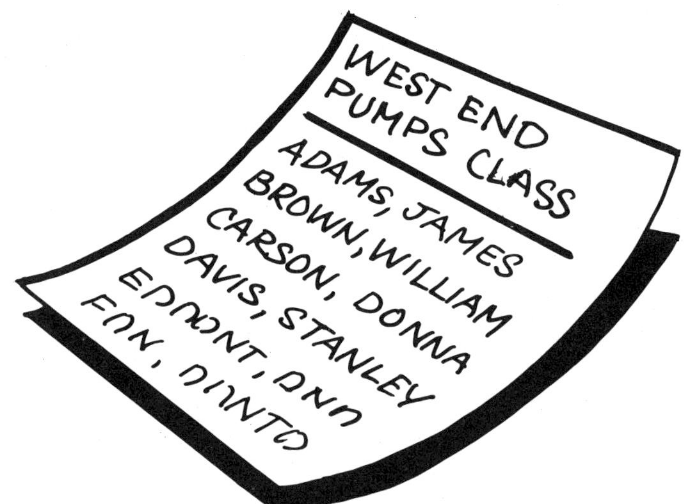

Fig. 1-12. "Classes" are people, not lists.

over a cup of coffee. Feel free to use "you," and "we," and "I." Be yourself. Teaching requires making true contact with people, exchanging information and ideas, sharing and generating excitement and enthusiasm, not just talking in their presence.

The Voice: Your Most Important Teaching Tool

Control *of* your students and communication *with* your students will be carried out principally by voice. We exercise control through instructions and commands, leading students in the directions and paths we want them to follow. We use the voice to counsel, to correct, and to impose discipline, guiding our people and keeping them within the boundaries established for the training program and by departmental policy. The process of teaching flows naturally from the foundations and techniques of class control.

While the fire service commitment to instructional media seems to double and redouble with each decade, there is still no substitute for a caring human being at the front of the classroom. There is no question but that visual aids will positively enhance much of our teaching. Certain subjects and certain segments of instruction lend themselves to various "teaching machine" formats such as slide tape, videotape, or computer-assisted instruction. We are convinced, however, that the best "teaching machine" is a

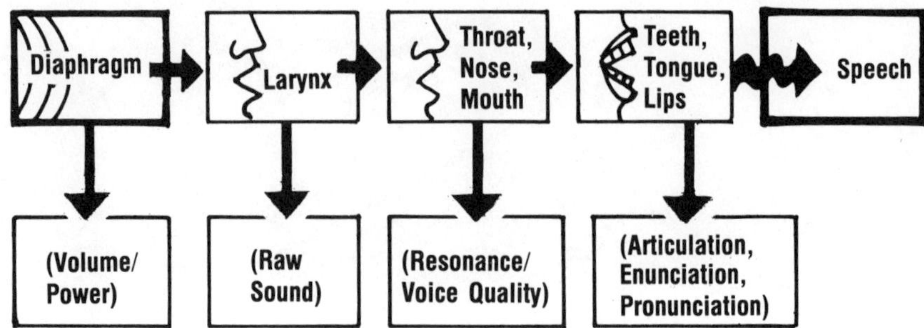

Fig. 1-13. Components of speech.

human instructor, using the most complex computer yet available, the human mind. The most fortunate students are not those who are fed by the latest in technology, but those who are led by an individual who is personally motivated and who is a motivator of others, standing eye to eye with the class, expounding, questioning, and challenging with the oldest and best instructional tool—the human voice.

Your voice is unique, as individual and as identifiable as your fingerprints. In teaching, we want to retain the qualities which cause your voice to be "you," while at the same time ensuring the best possible vocal and verbal communication.

We speak by using the diaphragm to push a flow of air across the "vocal cords" or larynx. The sound created, the unique sound of *your* voice, is a product of your vocal cord system and the size and shape of the air chambers existing in your throat, nose, and mouth. (When we say a certain voice has a "nasal" sound or is "throaty," we are acknowledging the effect of these spaces on voice quality.) Finally, this raw sound is converted into specific and understandable syllables and words with the teeth, tongue, and lips. We can directly control voice volume with the diaphragm, and we can alter the clarity of our speech by articulating and enunciating carefully. Voice quality and character, on the other hand, established by the larynx, throat, nose, and mouth, are relatively fixed and can be changed only through extensive and rather demanding training (Fig. 1-13).

Using the voice effectively includes varying the volume and rate of your delivery. A monotone pattern of speech is deadly dull, and you must, particularly if your voice tends to be flat and unemotional, seek ways to put some sparkle into your words. Pauses, for example, provide effective verbal "punctuation." Letting a statement "hang" in the air is a valid way to emphasize it. You can also use pitch to punctuate, making your voice rise to pose a question and lowering the pitch to state an authoritative point.

Fig. 1-14. Voice projection.

Increasing or decreasing volume will also provide emphasis for critical points you wish to stress. A loud word or phrase obviously will focus student attention, but don't overlook the advantage of reducing voice volume, sometimes almost to a whisper, as an equally useful way of stressing a point. Practice both extremes, concentrating your attention and your effort to improve proficiency where it is most needed. Few voices are too loud; many are too soft. If your voice tends toward the soft end of the scale, consciously practice using the diaphragm to *push* your words to the last row of students or to the back wall of the room (Fig. 1-14). Forceful speaking *can* be learned; volume *can* be increased. Remember, if they can't hear you, they can't learn.

Change pace, occasionally, to add further variety to your speech. Your personal baseline speed of delivery is usually a product of the region in which you grew up; Easterners and Northerners tending, of course, to speak more rapidly than Southerners and Westerners. One caution: regardless of the speech patterns common to your area, do *not* speak so rapidly that your students cannot follow. Keep ideas, points, thoughts, and principles flowing at a rate appropriate to the material you are covering and suitable for the abilities of your students. An old saying reminds us: "Too much, too fast, will lose the class."

Students need not only to hear your words; they must be able to understand. Be sure you speak clearly, articulating word sounds properly, enunciating plainly, pronouncing words correctly. We need not—in fact, probably should not—sound like professional orators, but we must be fully understood. Communication does not occur until the receiver comprehends the message. Proper volume, proper pacing, clear speech, and reasonably good grammar are all essential to ensure student comprehension.

We also communicate positively in a variety of non-verbal ways. Gestures are useful and add emphasis and vitality to your spoken message. Don't be afraid to use your hands to explain or emphasize a point. Facial expressions often say as much or more than your words and can, at times, totally change the message. Body language also speaks volumes. Strong voice vitality and strong physical vitality are both absolutely essential components in effective teaching.

Distracting Mannerisms

Watch for and eliminate distracting repetitions in your speech. The most common, and the most annoying, are "ah," "OK," and "ya know." When your students sit there counting the number of "OK's" you use during an hour of teaching, you *know* you are in trouble. Watch, too, for negative actions like constant pacing, jingling keys or change, tossing a piece of chalk, or waving a pointer. You should not smoke, drink, or eat during class. (Don't laugh—they all happen.) Your rule should be: "If it distracts students, don't do it."

One closing thought on distractions: One of our occupational hazards as fire service instructors is the radio monitor. It is impossible to communicate with a group of firefighters while a monitor is sending out an emergency call—*any* emergency call, theirs or not. Student attention will automatically "zero in" on the words of the dispatcher, and *your* words, along with any teaching point you were attempting to make, will be lost (Fig. 1-15). Monitors should be turned off in the training area. If they cannot, learn not to compete with them; you will lose.

Eye Contact

We communicate much more with our eyes than we realize. Our eyes have been referred to as "the windows of the soul," and so they are. We hear phrases like "a frosty stare," or "a loving glance," and we immediately gain a sense of the interpersonal relationships conveyed. We use our eyes to help us build our communication bridge with individual students, drawing them into one-to-one contact with us as our eyes meet theirs.

We are not going to lay down very many flat, dogmatic rules. Many of the thoughts we discuss in this book are ideas or techniques that we have seen work, methods that work for us, consensus opinions of instructors with whom we have worked, concepts and suggestions which we urge you to try

Fig. 1-15. Eliminate distractions.

for yourself, accepting those which fit your situation and rejecting those which do not. Eye contact does *not* fit that description. In small-group instruction, which is the usual situation for most fire service classes, eye contact is absolutely essential (Fig. 1-16). You *must* master the technique if you expect to become a first-rate instructor. To most of us, it does not come naturally. Effective eye contact is, however, a teaching skill which can be practiced and, through practice, learned.

As you face your class, consciously focus on the eyes of individual students, moving *your* eyes relatively slowly but regularly throughout the group from side to side, front to back, in whatever pattern you choose to ensure that, as you teach, you make definite, periodic contact with each person in the class. Obviously, if you are asked a question or are making a specific point with an individual student, you will focus on that person for as long as necessary. The goal is to pull each student into the mainstream of your teaching by frequently giving each one your personal attention with your eyes while, at the same time, you continue to speak to the entire group. There is no specific set pattern or particular cycle of time to strive for. We suggest only that if you were asked the question, "Did you look each student directly in the eye at least once during that last segment of instruction?" your answer should be, "Yep, I sure did."

One of the common pitfalls to be avoided is failing to make eye contact

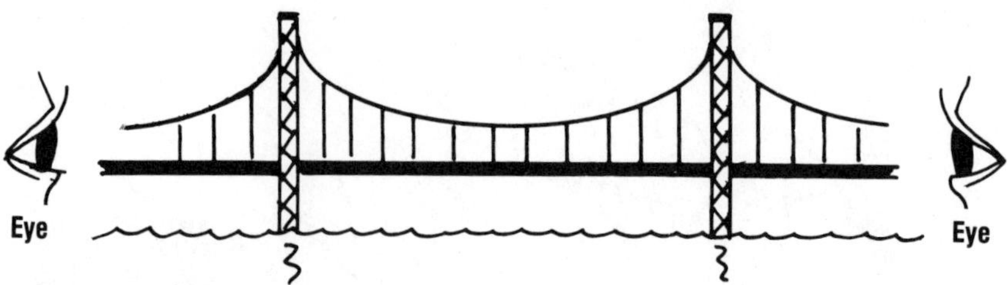

Fig. 1-16. Our communication bridge.

with the students to your far right and far left near the front of the room. They are the easiest to miss because they are on the periphery of your vision, and it is natural to "center" your attention toward the middle of the group. Also, most of us have a tendency to direct our attention and our teaching toward the students who appear to be closely following, those who are giving us positive feedback by nodding in agreement or otherwise giving a clear indication that they are listening and approving. Our egos feed on their obvious appreciation of our superior wisdom and wit, and we tend, like a nightclub performer, to "play" to that part of the room which is responding favorably to our performance. Even though it is difficult, try to avoid directing all your attention to the nodders.

One strange fact you will discover in teaching is that often the person who was nodding *so* positively did not absorb one single word you said, while some student on the other side of the room, who did not appear to be paying the slightest attention, caught, digested, and accurately remembers every point you covered. It can be positively weird.

Look Alert—Be Alert

Even though you will occasionally be fooled by the nodders, try constantly to assess the level of communication you are achieving. Questions you ask of the students and their participation in guided discussion will provide an easily obtainable and reasonably accurate indication of how effective you are. Think about the answers you are receiving. Constantly test yourself: "Do my students understand?" "Are they following me?"

Be quick to notice and "reel in" inattentive students. Pull them into the flow of your teaching with eye contact, a question, or by moving into their part of the room. Move about freely (Fig. 1-17). You are *"the boss"* and you have a perfect right to go where you choose to go, to stand where you want to stand. One of the most effective ways to break up a private, unrelated discussion between students, or to take charge of some minor disturbance in

Fig. 1-17. Go where you want to go; stand where you want to stand.

the room, is to move directly to the spot where the problem is occurring, continuing to teach as you have been teaching, continuing to make the point you were making, but also clearly indicating that you know what is going on and that *you* are *in charge*. No other action is usually necessary.

Part of the secret of successful class control lies in being alert, knowing what is occurring in every part of the room at all times. The rest of the secret is in *appearing* to be alert, making your students *believe* that you know what is going on in every part of the room at all times. Look alert and be alert. It isn't hard to do, and it really works.

"The Show Must Go On"

Speaking of appearances, it is often necessary to teach when that is the last thing you feel like doing. Minor illness, personal burdens which preoccupy your mind, work problems, all sorts of difficulties can make you feel like throwing in the towel. With a class scheduled and fifteen, twenty, or twenty-five students ready and waiting, there is little you can do but get on with it. You must, at times, act out a role you really don't feel like playing. If you can't *be* enthusiastic, you must do your best to *appear to be* enthusiastic. You will also find that displaying enthusiasm the twenty-seventh time you teach a particular class can require some *real* acting talent. Give it your best effort.

Fig. 1-18. Pride! Enthusiasm!

Instructor Characteristics

If you are sincerely interested in becoming a fire service instructor, we suspect you already possess many of the positive characteristics we have mentioned in the preceding paragraphs. Fire service instruction is not a profession people enter if they are looking for fame and fortune. Most of the folks in our business take pride in what they are; the badge and the uniform become symbols of more than mere departmental affiliation, they signify membership in the larger fraternity to which all firefighters belong. By far the majority of instructors with whom we have worked over the years *do* take pride in being a part of the fire service, and their enthusiasm freely extends to everyone they contact (Fig. 1-18). There is no better way to grow in any profession than to learn from those people you recognize as strong, effective, and worthy of your admiration. Pattern your instructional career development using the best people you can find as role models.

Strive also at every opportunity to develop your leadership ability to its highest possible potential. Leadership, supervision, and teaching are almost impossible to separate; the positive qualities of effective teachers and effective leaders are all cut from the same piece of cloth (Fig.1-19). Every leadership technique you master will make you a better instructor. Each new level of proficiency you achieve as a teacher will make you a better leader. Learning to organize effectively, illustrate appropriately, and present your thoughts and ideas smoothly as an instructor is anything but a dead-end career task. What better training could there be for the chief officer who must effectively organize, appropriately illustrate, and smoothly present the plans and programs of the department before the city council or before the board of trustees?

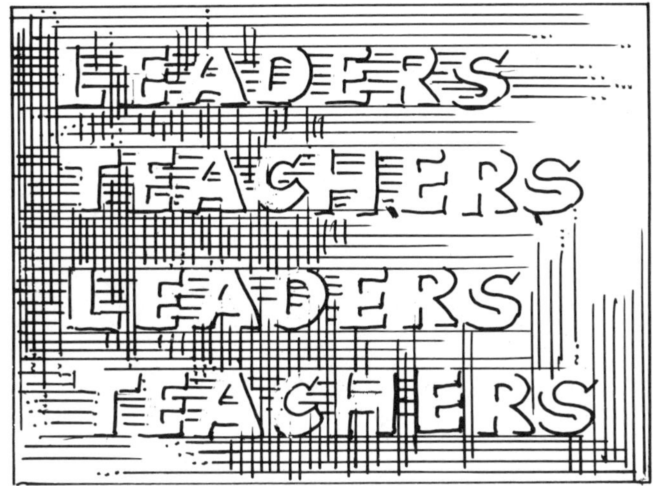

Fig. 1-19. Teachers and leaders, the "same piece of cloth."

Continue Always to Improve

The instructor who stops striving to improve can move only in one direction: backward. It is impossible to stand completely still in any profession, particularly in a profession like teaching which by its very nature must be in tune with tomorrow. You simply cannot rest on yesterday's achievements, no matter how well they were received. If you try to stand still, the explosion of knowledge, which is all too frighteningly real, the increasing sophistication of the machines at our disposal, and the changing technology of our profession will quickly march off into the future and leave you standing at the side of the road, wondering where everybody went.

You need to know what constitutes good instruction, to know what separates a totally superior teaching job from one which is merely acceptable. We can, we hope, look together at some of the factors that will tip the scale of *your* performance toward the very best instruction possible.

DetailsDetailsDetailsGoodDifferenceGreatDetailsDetailsDetails

As with most crafts, the difference between "good" and "great" instruction is hidden in a multitude of details. We will include a great number of these small but critical points. Many of them may not seem especially important in themselves, but employed collectively they will enable you to raise your personal level of teaching proficiency significantly, assisting you to communicate more effectively, use class time more profitably, or simply make your work easier and more fun. Genius has been defined as an infinite capacity for taking pains. Don't be too quick to throw aside some little suggestion just because it seems trivial or "not worth the effort." Every painting is made up of a multitude of little brush strokes; every stroke has a purpose, and every one is important to the whole. A superior teaching job is also made up of a multitude of little details; every one has a purpose and every one is important to the whole.

Learn from Others . . .

Productively use every opportunity you have to observe other instructors. As you participate in any class or as you listen to any speaker, think about what the teacher or speaker is doing, consider why it is being done, and assess whether or not it is working. One of the hazards in really examining instructional techniques, as we will, together, is that the process will make it almost impossible for you to sit in someone else's class without evaluating the performance. Learn from those performances. Every experienced instructor utilizes styles and ideas and techniques acquired from people he or she has observed and admires. Adopt the things that work; avoid those which do not.

. . . But Be Yourself

One note of caution: not *everything* works for *everybody*. Some of us tend to be rather formal; others are more casual in style. Some of us have a knack for blending humor into our presentations; others do not. Learn from others, borrow freely from others, but "Don't be what you ain't."

Being yourself is extremely important. One fear we have, with the emphasis today on uniform standards, is that we may destroy individuality and create a generation of "cookie cutter" instructors, all stamped from the same mold, all marching to the same drum. We encourage individual freedom, but, along with the freedom to "do your own thing," goes a responsibility to prove yourself as a successful instructor. Some of the greatest teachers we know in this business are real mavericks, people who break every rule in the book and yet can motivate students, capture their interest, and teach with a capital "T." Their license to break the rules is *proven ability*. "Being yourself" does not relieve you of the responsibility to follow the rules until you have proven you can carve your own path, and do it well (Fig. 1-20). We once read of a master artist who, when asked by a young painter to evaluate a piece of abstract art, said, "First, show me your drawing of a

Fig. 1-20. Rules first . . . freedom second.

human hand; then, when you have proven that you can draw reality, I will critique your abstract work." Rules first—freedom second.

Analyze your own characteristics. Each of us knows better than anyone else where our own strengths and weaknesses lie. Each of us, if we are completely honest with ourselves, can be his or her own best critic. Concentrate on the areas where you need improvement. We feel it is usually best to work on improving one element at a time. If you feel you need to improve your eye contact and also to improve your ability to speak more slowly and precisely, we suggest you tackle first one and then the other. Decide where to place your concentration and master that aspect of your delivery, or at least become comfortable with it, before beginning to work on the next task.

Profanity

This seems as good a place as any to mention profanity in teaching. Over the years we have heard it argued that an instructor "can't talk to firefighters without cussin'," or "My students expect me to swear." Baloney. In class it is unprofessional and unnecessary. A great many people are offended by foul language, probably more people than most of us realize (Fig. 1-21). Controlling profanity will never cause you to lose any points; not controlling it can, at times, get you into a lot of unnecessary trouble. Give the subject some serious thought.

If You Need Help, Ask for It!

From time to time, all of us run into problem areas in teaching. It may

Fig. 1-21. No, Non, Nein, Nyet!

be a piece of material which looks wonderful in our teaching notes but falls flat in the classroom. It may be a difficulty with a particular teaching technique or in the use of some particular training aid. If you find yourself struggling with something that simply will *not* work, ask another instructor to sit in on one of your classes. A fresh viewpoint on the problem from a knowledgeable colleague can often help you determine why things are going wrong and, if you are lucky, may also point to a solution.

A Gem of a Profession

Many years ago, in a keynote speech at the FDIC in Memphis, a grand old man, the late Joe Fetters, said some exceedingly important words about fire service instruction when he admonished us to ". . . search for and polish new facets in this gem of all professions." Within the fire service, which we all love, the role of the instructor is most certainly a "gem" of a profession and one well worth polishing (Fig. 1-22).

Each of us, as instructors, will leave a legacy of unimaginable worth. Many of you younger people will, late in your careers, teach firefighters who are not yet born as you read this page. That is a certainty. By the time you come face to face with this yet unborn generation, we, most probably, will no longer be here on this earth. But *we* can touch *them* through *you*. That idea is a tremendously exciting prospect. We have wondered: "Who were the instructors who taught the instructors who taught us?" When we stand here, on the brink of retirement, and link the chain of people who taught us to the chain of people *you* will teach, the lifetimes of the fire service people

Fig. 1-22. A gem of a profession.

involved extend from the closing years of the nineteenth century to well past the middle of the twenty-first century (Fig. 1-23). It is a rare privilege to be a small part of a work which spans nearly two centuries, one which is so vitally important to the lives and the safety of so many human beings, most of whom we will never see or know.

A Brief Look at Our Students

Like their instructors, fire service students come in all shapes and sizes and bring with them an infinite variety of human diversity and needs. Like us, they also share some common characteristics. They are, in general, mature and ready to learn. They are keenly interested in practical applications and in learning "how" and "why" things work. Most of them have serious reasons for being in our classes, and they are quick to respond to, and appreciate, competent instruction. That very positive note is a plus for our side, but there is an obvious corollary: our students will be just as quick to detect an instructor who is incompetent or unprepared.

While they are similar in their generally serious desire to learn, our students differ from one another in many respects. You will find in your classes individuals who are diverse in age, education, experience, intelligence, and physical ability. Volunteer firefighters, in particular, may be found at virtually every point on the curve which could be drawn for each of these characteristics—from quite young to very old, from folks who are

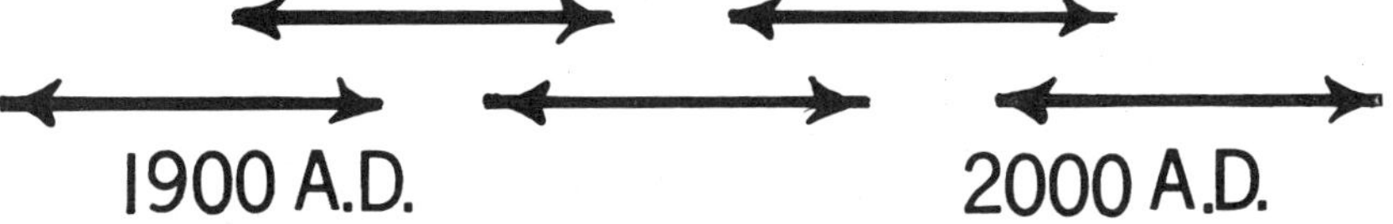

Fig. 1-23. Our influence is part of a lengthy chain.

illiterate to scholars with advanced degrees, from those who joined yesterday to people with a lifetime of experience, from bright to dull, and from those who are physically strong and adept to those who are incapable of satisfactorily performing parts of the required tasks. Some volunteer organizations establish minimum standards for active membership; others do not. While the techniques and equipment used today in firefighting require personnel with relatively high levels of mental and physical proficiency, we are inclined to feel that, within volunteer organizations, it is also important to give some appropriate role to all who wish to serve. It is certainly not our purpose to discourage minimum standards. We merely suggest that our attention to standards be tempered with consideration for people who have a desire to serve but who perhaps cannot effectively function as members of the first-line interior attack team. We feel it is part of our responsibility to help them find other roles they can fill.

Instructors working in career fire departments will also be faced with student differences, but the range, the distance between the low and high limits, for the characteristics mentioned above will be significantly smaller. Rigid minimum entry standards, the extensive initial training offered in most departments before assignment to an operational company, and pension systems which permit retirement at a relatively early age all work to reduce differences among paid personnel.

Every fire service instructor is plagued by this problem of divergence within the student group. For the novice instructor it is particularly frightening and frustrating. We stare at the lesson plan and ponder, "How in the world can I make this block of material meaningful and useful for the newest people on the job without completely and totally boring the old hands?" Fortunately there is a statistical certainty that comes to our rescue.

Central Tendency

The differences we encounter among our students will be governed by the principle of central tendency (Fig. 1-24). This concept reminds us that, within a given group of people, any measurable characteristic—age, for example—will tend statistically to follow a "bell-shaped" curve. A few will be very young, a few will be very old, and by far the majority will pile up in the center, or mid-range, around the "average" age. We teach toward the center of the curve, not ignoring the students who may be very slow or very quick, but planning and teaching our lessons for the majority. There are techniques for effectively bringing those at both ends of the scale along with the rest of the class, and we will touch on these from time to time in later sections.

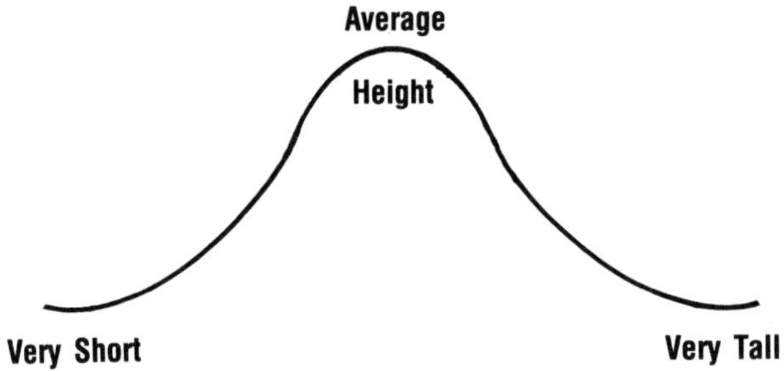

Fig. 1-24. Central tendency.

Teaching Adults

One characteristic our students *all* share is that they are not children. In recent years there has been increasing interest in adult learning as it differs from the general science of teaching. We do not have the space or the expertise to explore the subject in depth, but it is fairly obvious that children *are* different. Their "mental slates" are clean and ready to be filled. They have not had time to build up biases or prejudices or to have acquired reasons why "It can't work that way." Their minds are like little sponges, soaking up everything in sight, and they are, in general, more likely to believe whatever the teacher tells them.

Adults are almost diametrically opposite. Their "mental slates" are full to overflowing. They come to class with all sorts of preconceived notions and ideas, loaded down with experience which makes them skeptical, biased, and often downright hostile to many new concepts. They constantly measure the words of the instructor against what they already know, or believe, to be true. They are a critical audience.

Furthermore, adults bring to the classroom an abundance of other barriers to learning. Every one of them is carrying a full complement of personal interests, and each will have some measure of individual and private problems occupying mind and attention. They have a strong tendency to let their minds drift off to their own concerns, particularly if the instructional presentation does not succeed in holding their attention.

One asset adults bring to us is a generally high ability to learn. Human ability to learn climbs rapidly during our early years, accelerating in leaps and bounds as new skills are acquired, and probably levels off at about age twenty-five or so. If there is a decline from that point, it is probably more a matter of attitude than physiology. The "average" individual at age fifty or sixty is still capable of learning at a high level provided he or she is interested and motivated. We recognize, of course, that there is no "average" individual, that learning abilities will differ. The important point is to put out of your mind the idea that, "You can't teach an old dog new tricks." You *can* and, in fact, you *must*.

Teaching New Tricks to Old Dogs

Adults normally must be made to see some practical application for the information you are teaching. They will learn, in general, only that which they feel a *need* to learn. You should be ready to provide answers to such questions as: "How is this subject going to help me as a firefighter or fire officer?" and "Of what practical use or value is this information?" Even if your students do not ask such questions openly, you can be sure they will be *thinking* about them, and it is certainly to your advantage to convince and assure them of the utility of the subject material. Someone once said that instructing adults is like working with a piece of chain. You cannot push it, but if you take the lead and gently pull it along with you, it will follow you wherever you want to go.

Adult learning most often centers on problems, and the problems we pose in class must be as realistic as possible. An excellent technique in gaining adult student interest and attention is to tie the topic into a pertinent, real-world experience with which the class is familiar. The closer your example hits home, the more powerful the motivational effect. Always use specific cases to lead into theory, not the other way around. In other words, do not teach the theory first and then show a practical application. Capture students' interest with a practical problem they clearly understand and then teach the theory which applies to that problem.

Don't be afraid to repeat key elements; repetition reinforces adult learning. Using current instructional terminology, we give students objectives, then make the presentation, and then give them a summary. A most effective old-fashioned backwoods preacher once said, "I tells 'em what I'm gonna tell 'em. Then I tells 'em. Then I tells 'em what I told 'em." Same thing, exactly.

Adults usually learn best by doing. The more they are actively involved, mentally and physically, the better. Provide opportunities to discuss, to argue, to practice, to *do*. Provide opportunities to apply the concepts you are teaching. Activity heightens interest and helps block out the other thoughts and worries which weaken your hold on their attention.

As a rule, adults learn best in an informal environment. If it can be accomplished without sacrificing a reasonable level of discipline and control, students should be made to feel "at home" and "relaxed." The degree of informality you permit or encourage will depend on departmental policy, on your personal preference, and, of course, on the size and level of the group.

Adults are more commonly interested in guidance and personal improvement than they are in grades. They will seek your opinion on, and approval of, their progress. You will hear: "How am I doing?" and "Have I got this right?" and "Is this what you want?" A word of encouragement, a pat on the back from you will be far more valuable than you can believe possible. This subject, too, is a topic we will return to later.

"Problem" Students

We do not feel the "problem" student is as common in fire service classes or as much of a difficulty as some instructor training courses would lead us to believe. On those rare occasions when you are faced with a significant student problem, your first and best reaction should be to arrange a private conversation with the individual involved, outside of class time and away from other members of the class. Attempt to determine the specific reason *behind* the problem. Assure the student of your positive interest and your desire to be of assistance. Be available, but do not pry. "Counseling" is a currently fashionable word in fire service training but, if really serious personal problems are involved, few of us have the experience or training to provide appropriate assistance. True counseling requires a highly trained and skilled counselor. Be careful. Limit your "counseling" to areas where you are on firm ground and where you are sure of what you are doing.

If a student or a group of students disrupts your class to the point where you cannot control the situation, work through the existing chain of command to get the offenders into line or out of the classroom. It is not fair to you and it is certainly not fair to the other students for a small minority to destroy a learning opportunity. Don't let it happen.

. . . An Encouraging Word

As we have said, several times and in several different ways, we believe fire service people are rather special. Our business does not encourage "prima donnas" or any other form of antisocial behavior. People who don't fit are quickly *sorted out* or *straightened out*, either by their peers or by coming face to face with the deadly realities of firefighting. You can teach firefighters for a long, long time without meeting a serious personnel problem in the classroom. Your successes in dealing with your fire service students will far outnumber the few times you will fail, and the rewards received, in terms of personal satisfaction for both you and the people you teach, will far outweigh the cost.

Well, that's a thumbnail introduction to the players in our game: instructors and students. Bear in mind that human beings are highly resistant to being categorized or placed in preconceived molds of any sort. We've used a lot of hedging—words like "many" and "most" and "in general," not because our points and principles are of questionable validity, but because, for any rule of human behavior there will most surely be at least *one* exception. The exception just might turn up in *your* class. Consider yourself warned.

Now that we've introduced the players, come along and we'll look at the playing field. While we're walking over to the academy, we'll tell you about some of the problems in our profession that really upset us. The situations in which some of our people are expected to teach are absolutely outrageous.

2

The Learning Environment

Let's face it—it *is* possible both to teach and to learn under completely outrageous conditions. Fire service instructors and fire service students have provided proof of that, over and over again, for far too long. It's about time we challenged those conditions and said, "No more!"

Only in relatively recent years have fire stations been designed and built with classrooms included, some with absolutely first-rate educational space and some with dual-purpose areas which can reasonably serve the purpose. Even with this modest progress, today it is still not unusual to see classes huddled around the back step of an engine, with poor lighting, makeshift seating, and no decent work space for either the instructor or the students. We find classes held in the firehouse recreation room with the instructor working from the pool table while students lounge around the perimeter of the room in overstuffed chairs, or in the kitchen, or in space shared with majorette practice, or the neighborhood aerobic dancing class, or the television set, or something equally distracting (Fig. 2-1). It's all been tried at one time or another.

Critical Bearing

The fact that it has been done before, however, does not make it right.

Fig. 2-1. Chaos.

The learning environment has a critical bearing on the effectiveness of your instruction and on the ability of your students to learn. The finest lesson plan, the best teaching technique, will be diminished, if not totally destroyed, by poor physical arrangements. The most highly motivated students cannot learn if they cannot hear the instructor or if they are unable to concentrate on the instruction. Since we, as instructors, are held accountable by our management for teaching others, we have a right to expect a suitable and appropriate place in which to teach, and we have a responsibility to our students to demand a suitable and appropriate place in which they can learn.

Freedom from Distraction

You cannot teach, and your students cannot learn, in an atmosphere filled with noise and confusion. Among the most common problems in "makeshift" fire station teaching situations are radio monitors going full blast, apparatus engine noise, and personnel coming and going through the "teaching space" on their way to the soda machines, the station office, or some other part of the building.

Noise and traffic through the area can often be eliminated or greatly reduced by requesting the cooperation of personnel or, if necessary, by order

Fig. 2-2. Calm.

of the officer in charge. More often than not, our problems are a result of thoughtlessness or a lack of understanding, rather than a deliberate attempt to interfere with the class. If you are teaching away from your home facility, visit the proposed class site in advance, if at all possible. Whether at home or away, note any foreseeable items which might disrupt your work and discuss both the problems and your suggested solutions with the station or facility officer. It is often possible to turn monitor speakers down or off in class areas. Personnel can be instructed to stay out of that part of the station while class is in session, and apparatus engines can be run before or after class, or away from the teaching space.

To Teach or Not to Teach

There should be an absolute commitment from the officer in charge that any regularly scheduled teaching session will have priority over *all* other activities and functions, except emergency calls (Fig. 2-2). You should not have to plead or argue repeatedly for reasonable peace and quiet. If you make a positive effort to state your needs and if you believe those needs are clearly understood, we feel strongly that you should stop teaching—simply shut the class down—if you find your work, or the work of your students, regularly disrupted to any serious extent. Be reasonable, of course, but be firm. No lawyer is expected to plead a case with disorder in the courtroom; no actor is expected to perform with unwarranted noise in the theater. Why, then, should *you* be given any less courtesy?

Altering or Designing Teaching Spaces

Some of you will be working in excellent classroom facilities, some of you in teaching spaces which are marginal, and others of you will be faced with some of the truly deplorable conditions we have just described. The range of your needs in this area, therefore, will be quite wide. We will present the various classroom features we think are important, *without* locking our discussion specifically to either alteration or design. As your needs dictate, you can apply these suggestions and ideas to the upgrading and modification of existing multi-purpose spaces, or use them as a guide for planning purpose-built classrooms.

Let There Be Light

Adequate lighting is essential. A proper classroom will be designed with an appropriate level of lighting, sufficient to ensure that reading, writing, and any other necessary activities can be performed with absolutely no eyestrain. Ideally, lighting levels should be variable by means of dimmer controls or a bank of individual lighting switches, located near the instructor's normal teaching position. This location permits the instructor to choose illumination levels and arrangements appropriate for specific teaching-learning situations, and it enhances the display of slides, overhead transparencies, and similar projected media. Quite frequently, a low lighting level, one which permits students to take notes and gives the instructor a chance to maintain eye contact with the class, is preferable to complete darkness.

If the area assigned as teaching space does not appear to meet reasonable minimum standards, consider increasing the wattage of bulbs, washing the lighting fixtures, or both. If these changes do not give you the light you need, we suggest moving the class to a more appropriate location. Neither you nor your students should be expected to function in perpetual twilight.

Heating, Ventilation, and Cooling

Reasonable temperature levels are required to maintain student comfort. The specific temperature depends on the activity; students sitting at tables taking notes during a lecture may feel chilly at a temperature which would be perfectly comfortable for performing skills. The "comfort range" for classroom teaching is relatively narrow, probably somewhere between 65 and 72 degrees Fahrenheit. Listen to your students. If they are not complaining, you can assume they are reasonably comfortable. If they are not comfortable, they will usually be quick to let you know.

Dedicated classrooms should have individual HVAC systems, giving the

Fig. 2-3. No smoking in classrooms.

instructor control over temperature and ventilation. Keep some fresh air moving, if at all possible. A stuffy, overly warm room will put your group to sleep faster than anything else, particularly in the hour right after lunch.

Most adult educational institutions today prohibit smoking in classrooms (Fig. 2-3). Unless local custom is totally contrary, neither students nor instructors should smoke during indoor classes. On the same principle, it is generally considered good practice to require that food and drink be kept out of the classroom. This prohibition eliminates an unnecessary distraction and avoids the problem of spills and sticky tables.

Seating and Table Space

Student seating should provide both comfort and an attentive posture. Overstuffed armchairs or couches are too soft and encourage lounging, while unpadded wood or metal chairs quickly become uncomfortable. Ample, uncrowded table space for student materials and note taking is absolutely essential. Desk armchairs do not provide enough work space and are generally inadequate for effective adult education.

We spend a lot of time in the fire service considering the "human engineering" aspects of apparatus cabs and pump panels. Take time to look at the "human engineering" of the student work station. Sit in the seat; work at the table or desk provided. Consider the lighting and temperature mentioned in preceding sections. If you are not completely at ease, the chances are that your students will not be at ease either. You, as their instructor, will be working under a major disadvantage. If it is necessary to fight for decent lighting, seating, and adequate table space, any effort you put into winning the battle will pay big dividends for your students and for you.

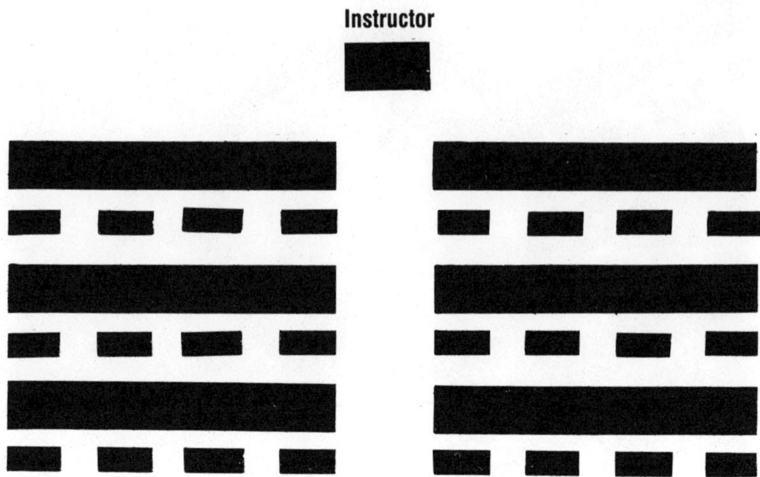

Fig. 2-4. Traditional row seating.

Seating Arrangements

A variety of seating arrangements are possible, each with advantages and disadvantages. Consider the nature of the class to be taught and the type of interaction desired. Interaction can be encouraged among the students, between the students and the instructor, or both, depending on how you arrange the classroom.

Traditional "Squared Away" Rows

Classic classroom row seating provides the maximum number of seats in any given space. This arrangement puts student attention straight ahead, focused on the instructor and on visual aids displayed at the front of the room. Row seating, therefore, works well with large groups in lecture and illustrated lecture situations. It permits instructor-student interchange but discourages group discussion because students cannot, without a lot of twisting and turning, see each other eye to eye (Fig. 2-4).

When arranging row seating, if possible, leave yourself a center aisle so you can move out among the students if you decide to do so. It is extremely helpful, many times, to get into close contact with the group. Moving out among your group helps build that communication bridge we talked about, helps to transmit clearly and positively the points we are trying to convey. For the same reason, *never* arrange a classroom with any substantial amount of

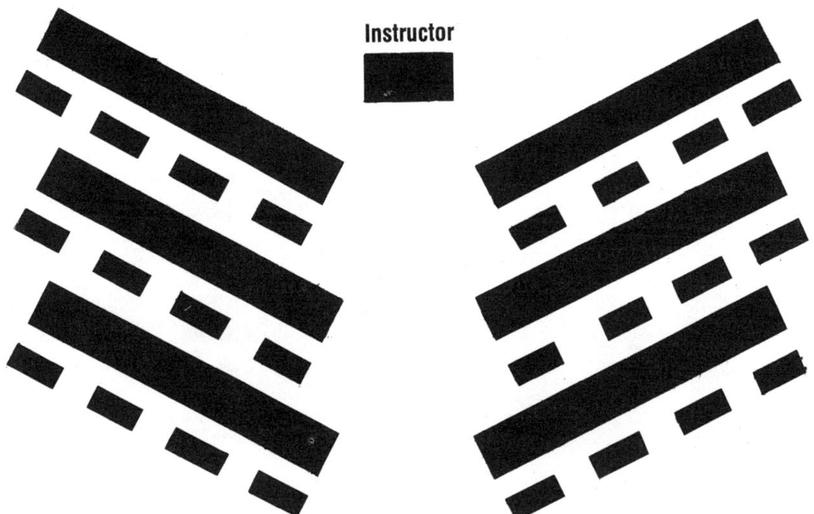

Fig. 2-5. "Chevron" or "herringbone" seating.

open space between you and the class. It is *extremely* difficult to establish good, solid communication across a significant gap. If students do not fill all the nearer seats and congregate toward the back of the room, move them to the front seats before you begin teaching. They may grumble, but it will make communication, and therefore your job, much easier.

"Herringbone" or "Chevron" Rows

Angled seating rows waste a little space but provide a more effective teaching arrangement. This layout creates a less formal, more congenial classroom atmosphere. Student attention is again focused on the instructor, but each student can now see the faces of at least part of the class. While this facilitates interchange between students, the instructor still maintains a high degree of control because attention centers more easily at the front and center of the room (Fig. 2-5). This arrangement is excellent for projected aids and will work for demonstrations in which close-up viewing by the students is not a critical factor.

The U Shape

The "U" arrangement diminishes the central focus on the instructor but provides excellent student interaction because most class participants can

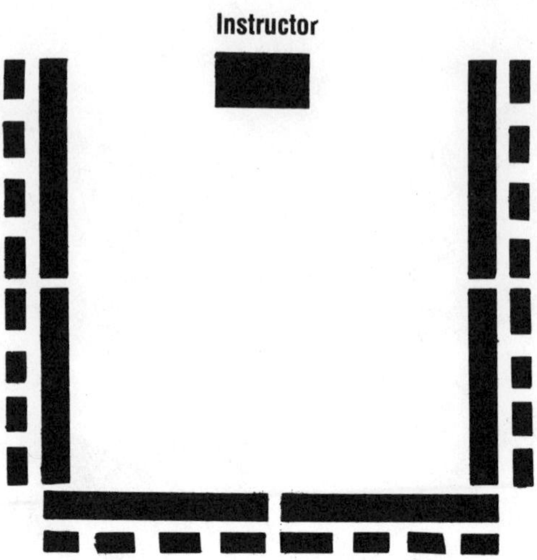

Fig. 2-6. "U" shape seating.

see each other eye to eye. Working at the open end of the "U," the instructor can move into the group to control a part of the discussion, or the instructor can withdraw, encouraging students to discuss or argue points more freely (Fig. 2-6).

The open end also provides a point at which the instructor can lecture or display visual aids. This position provides students across the bottom of the "U" with a clear line of sight but puts the students along the sides at a slight disadvantage. Just within the open end of the "U" is an excellent position to perform most demonstrations because *all* students have a clear view of this particular spot.

The "V" shape is a modification of the "U," a sort of compromise between "U" and "herringbone." The main reason we mention it is to encourage you to be innovative, to use your imagination in setting up the classroom. The main criteria for judging any seating arrangement is its utility in allowing you, the instructor, to accomplish what you want to accomplish. Be creative; don't get locked in to a fixed idea. Arrange the room in any way that you feel will give you and your students the best advantage in communicating effectively.

The Circle

For some classes, particularly certain types of skills instruction, the circle works really well. Students are seated or stand around the perimeter with one or more instructors working in the center. In this configuration, instructors can easily demonstrate a physical action, such as donning

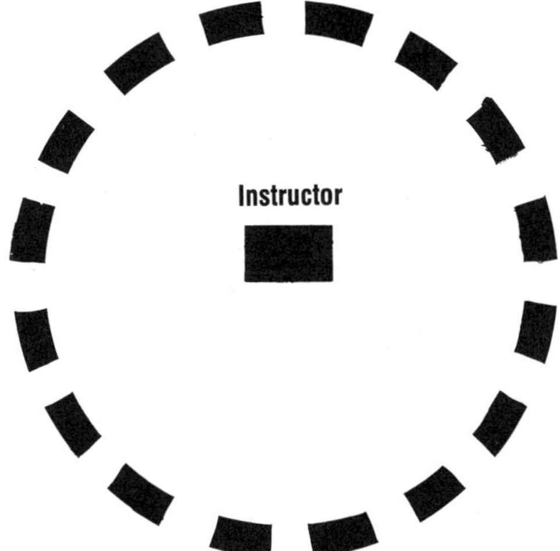

Fig. 2-7. The circle.

breathing apparatus, repeating the demonstration several times so all students see the action from several different perspectives (Fig. 2-7). When students begin to duplicate the skill, instructors can readily observe student performance and move quickly to those needing assistance or correction.

Some Thoughts Around the Perimeter of the Classroom

Chalkboards are available in a variety of colors, green and brown being the most common, in addition to the traditional blackboard. Magnetic-faced boards are becoming common, along with an excellent new writing surface consisting of coated white metal panels on which the instructor writes or draws with special, easily erased, felt-tipped pens which are available in a variety of colors.

In new installations, be sure board space is ample and be sure boards are installed at heights suitable for adult use. Many are installed too low, difficult for students to see and difficult for the instructor to use.

A permanently mounted pull-down or power-driven screen of the proper size should be provided in every dedicated classroom and, where possible, in each dual-purpose area regularly used for teaching. We will go into more detail concerning screens in Chapter 8.

Free-standing or table-top lecterns provide the instructor with a place to lay out teaching notes and furnish a more or less formal "home base" for structured parts of the teaching session, such as opening remarks or closing summaries. The working surface of the lectern should be of ample size.

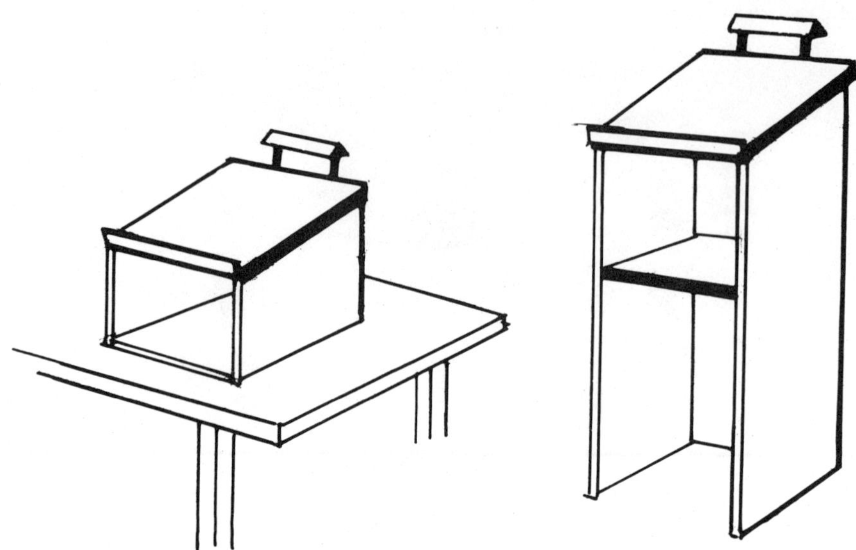

Fig. 2-8. Table-top lectern and free-standing lectern.

Lecterns should be constructed with open cabinet work on the instructor's side in order to provide space for storing small items. A low-intensity reading light and a pencil ledge are useful touches (Fig. 2-8).

Somewhere near the lectern you should have stashed away a pointer. We know this item is an insignificant little teaching tool compared with chalkboards and screens and lecterns, but it is *extremely* difficult to work "pointers" into a conversation. Bear with us; we'll only take a minute.

A traditional rubber-tipped wooden pointer, or one of those nifty little telescoping types, is useful for pointing out specific items on graphics or specific parts on objects or models. It focuses student attention exactly where you want it, and it gets your body out of the students' line of sight. Put the tip of the pointer on the spot you want to mark and *hold* it there while you are making your point. Don't just wave it in the general direction of the aid: that motion does nothing but distract. Speaking of distraction, when you are finished using the pointer, *put it down*. Recent educational studies prove, conclusively, that the average instructor *cannot* handle a pointer for more than 14.9 seconds without imitating either Arthur Fiedler or D'Artagnan (Fig. 2-9).

Electric pointers are essential for directing student attention to specific points on a projection screen, particularly when you are working with slides. They are supplied with plug-in cords, or with batteries, or both. The beam of the pointer may project a small arrow or only a dot of light. The newer

Fig. 2-9. "Put it down!"

laser-beam pointers are exceptionally bright and are "state of the art." As you do with stick pointers, put the beam where you want it, hold it still, and get it off the screen when you are finished. We offer the same advice as that given above: when you are finished, put it *out* and put it *down*.

"Blackout" window blinds of good quality are essential. Window darkening should include some means for maintaining ventilation if the room, at times, depends on outside air.

It is unusual to find a classroom with a sufficient number of properly located electrical outlets. They should at least be available at the front and back, centered, and, if the room is large, recessed in the middle of the floor—grounded circuits only, please. An outlet high in the center of the back wall provides the proper spot for a good-sized clock, facing the instructor and behind the students.

Cabinet or closet space is necessary for audiovisual equipment and other needed supplies. If at all possible, projection equipment and other AV gear should be permanently assigned to and remain in that room. These items will then be available when needed, stay clean longer, and be less subject to damage.

Finally, classroom color and decor should be subdued and unobtrusive. Students are a captive audience. Be sure the space in which they are held captive is reasonably pleasing and appropriate for learning. If finances permit, carpeting is a welcome addition in fire training classrooms where outside dirt will not be tracked in. Carpeting provides a quiet and comfortable walking surface, along with general sound absorption. Acoustics are ex-

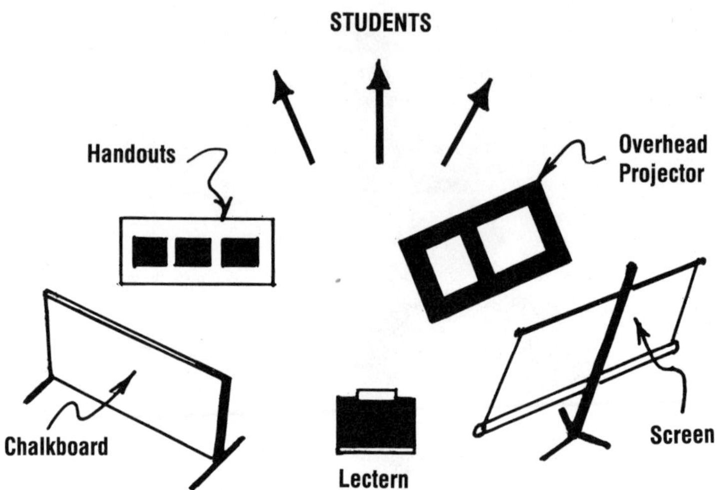

Fig. 2-10. Properly arranged teaching space.

tremely important and should be carefully considered in designing or converting classroom space. Totally hard-surfaced floors, ceilings, and walls create echoes; they distort words and make listening extremely difficult.

Speaking of sound transmission, we do not care for room dividers. They are awkward to use and they *all* transmit sound to some degree. Try to get additional purpose-built or dedicated rooms rather than attempting to divide one large space. Such division is seldom, if ever, satisfactory.

One Final Note

In setting up the classroom, pay particular attention to the areas *you* will be using. First and foremost is the area from which you do most of your teaching, usually at the front and center of the classroom arrangement. Within this area will be a number of key elements: the lectern, the chalkboard, the overhead projector with space for laying out transparencies, handout materials, display or demonstration equipment, and perhaps a video monitor, a flip chart, or some other special teaching tool—any and all of the items you want close at hand during class.

The important point to remember is that the physical arrangement of these key elements can either make your teaching job flow smoothly or it can be terribly disruptive. Think carefully about the layout (Fig. 2-10). Can you move from the lectern to the chalkboard without passing through the projection beam? Can you turn to write on the chalkboard or the flip chart in one swift motion? (Right-handed instructors will arrange the various items differently from left-handed instructors.) Are handouts and demonstration items at a convenient spot, within easy reach? Are all electrical cords placed so you can work without tripping over them?

Handouts or demo items can be arranged on the same table with the lectern, on a side table, or on a table between the lectern and the first row of students. There is no one correct or magic set-up. If an arrangement *works* for you, it is a good arrangement. The trick is to be sure you don't learn that you have an *unworkable* arrangement with twenty students looking on, thoroughly enjoying your misery as you slowly sink in a sea of misplaced teaching aids.

The second area to consider is the space from which you project films or slides, toward the back of the room in the center. Give yourself easy access to this spot. Organize student seating so you need not climb over or plow through students to get to your projection equipment. Maintain some clear space around the projection table or stand; give yourself room to work. Think about reserving yourself a seat near the projector if a film is lengthy and you want to stay in the room while the film is being shown.

Ensure that cords are placed so students won't trip, causing a fall or causing projection equipment to be pulled over. Taping cords to the floor, as much as possible out of the way of students, is worth the small extra effort it takes. A roll of duct tape, or "silver" tape, in your teaching kit will serve for this purpose, and it will come in handy for a lot of other things, such as mending a torn screen or temporarily whipping a rope. Duct tape sticks to everything.

Arrange a good firm base for your projectors. Solid placement will give you a better show and will help prevent damage to the equipment. Considering the price of these machines today, you should earnestly avoid the necessity of trying to explain to the chief how one fell three feet to a concrete deck, especially if you had balanced it on a board supported on the backs of two folding chairs because you were too lazy to go and get the projection stand. Trust us—it is *not* a good idea.

We mentioned human engineering a few sections back. Engineer *your* working spaces to meet *your* needs. Teaching is a difficult job. It requires strict attention to detail, an ability to think on your feet, and the capacity to react instantly and correctly in a complex web of dynamic interpersonal relationships. Anything which reduces your concentration on the teaching-learning situation should be eliminated if at all possible. The time and effort you put into carefully and painstakingly setting up your teaching and projection stations will pay off by enabling you to concentrate your attention on the needs of your students, not on the needs of their instructor.

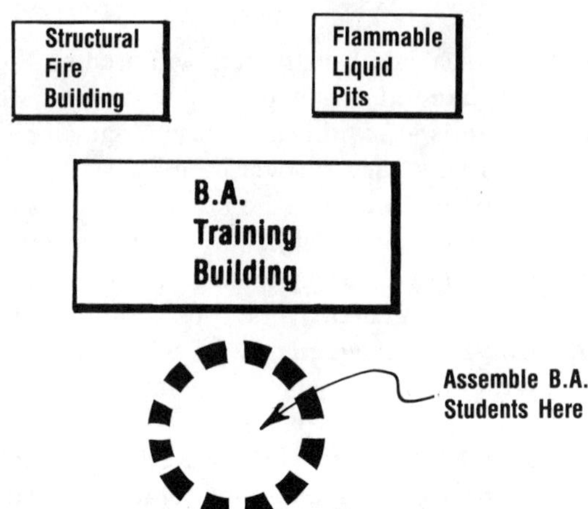

Fig. 2-11. Proper drill ground positioning.

You will learn more through experience than through anything we suggest. Once you have discovered what works and what does not work for you, it will become second nature to set up your teaching stations and to arrange student seating quickly and surely.

The Drill Ground

Teaching outdoors is an entirely different ball game. Maintaining group attention and group control are much more difficult. Away from the formality of the classroom, students relax and are more inclined to chatter and horse-play. The instructor often must contend with visible and audible distractions which sometimes surround the class on almost every side. Wind and weather interfere; your words, at times, are actually blown away. Noisy, cold, hot, wet, and dirty—teaching outdoors is fun, but it's not easy.

If a number of teaching activities are going on at the same time, position yourself, if you can, so your students are facing away from the most obvious distractions: live firefighting, for example. Many times you can move a group around the corner of a drill ground building, cut off both sight and sound distractions, and maybe, if you are lucky, move out of the wind as well (Fig. 2-11).

The student-to-instructor ratio must be much lower in teaching practical evolutions, and it is helpful to divide large classes into smaller groups. Small-group instruction gives us better control over class discipline and safety. It improves communication and provides the close supervision essential to effective skills instruction.

You may, at times, find a power megaphone or "bullhorn" helpful, particularly in getting a relatively large group organized and placed where you want them to be. Moving students from one area to another or from one drill ground prop to another is extremely time-consuming; people will use any excuse to straggle or wander off. Try, if possible, to incorporate moves into break times. Dismiss them at point "A" and have then reassemble in ten minutes or fifteen minutes exactly at point "B." It won't solve all your problems, but it will help.

Leave the Classroom Inside, Where It Belongs

In general, classroom teaching should be done in the classroom. Outside, charts blow away and projected images are hard to see. Projection equipment and other delicate visual aids are exposed to dust, dirt, and water. It is difficult for students to take notes outdoors; if the weather is windy or it is raining even slightly, note taking is impossible. Don't require it. Teach the principles; give the points you want noted in the classroom, before or after the drill ground activity.

If *you* need notes for outdoor teaching, putting your outline on a pack of three-by-five-inch cards works well, particularly if you cover each one with clear adhesive-backed plastic on both sides to keep them clean and make them last longer. Most important, without this plastic covering, the first spray from a hose line will leave you holding a handful of illegible mush. Slip a rubber band around the cards and drop them in the pocket of your turnout coat where they'll be ready when you need them.

Training centers in some parts of the country have carport-like spaces with open shed-type roofs over concrete slabs. These spaces are set up with folding chairs, to provide student assembly and instruction sites. Sometimes these structures include outdoor electrical outlets and are closed at one end with a chalkboard or a projection wall and sometimes side wings, four feet to eight feet in width, to shade the screen and to give the instructor a bit of additional shelter (Fig. 2-12). The utility of this arrangement depends to a large extent on local weather patterns.

Speaking of weather, if you regularly teach out of doors in a part of the country where weather conditions may force the cancellation of a class, have a back-up indoor lesson plan in your briefcase so the scheduled time can be used profitably. Don't waste a scheduled class period just because you cannot carry out the instruction originally planned.

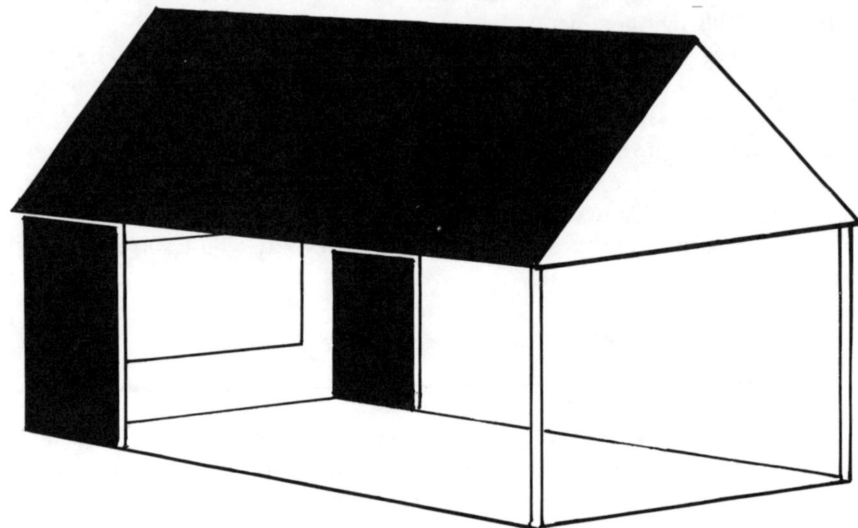

Fig. 2-12. Typical outdoor teaching shelter.

The Environment and Safety

We have talked about many environmental factors which can have positive or negative effects on teaching and learning. Before we leave the environment, we must also talk about safety. Let there be no question: The most negative thing which can occur in a training class is a major injury or a fatality. We cannot totally eliminate the possibility of an accident during training, but we can, through alterations in attitude and environment, greatly reduce the probability.

Perhaps "accident" is not a good word to use. Accidental implies "without cause," or "involuntary," or "beyond control." We recently read that there are no accidents, only caused occurrences. We agree, and therefore we must consider how these "caused occurrences" can be eliminated or minimized.

Attitude is fundamental. In the early 1900s, the chief of one of our major fire departments stated; ". . . firefighters are going to be killed right along." This philosophy, that injury and death are an inherent and unchangeable part of our work, has been handed down from one generation of firefighters to the next like holy writ. When a member dies, we salute, blink back the tears, and think, "Well, it's just part of the game." As a matter of fact, firefighters take a sort of perverse pride in the knowledge that ours is

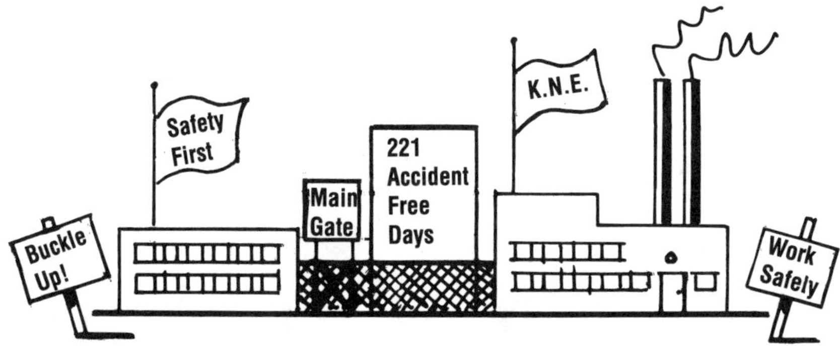

Fig. 2-13. Safety pays!

the most hazardous peacetime profession. It does not *have* to be, and the fire service instructor is in the best possible position to help change it.

Consider the Example of Industry

Industry decided a long time ago that safety is good business. Partly because of an interest in the welfare of their employees, primarily because safety is profitable. Accidents cost money. Before entering training, we served as fire chief in a United States Steel Corporation plant which set back-to-back industry safety records of 11,000,000 and then 15,000,000 man-hours without a lost-time injury. Those records were made possible because of the total commitment of the corporation to safety, because of a positive collective attitude. Every person in the mill, from the general superintendent down to the lowest floor sweeper, was involved in the safety effort. *That* is the level of commitment for which the fire service must strive. Safety will not work until everyone, from the top brass in the front office to the newest private on the back step, is fully convinced and totally involved (Fig. 2-13).

The primary U.S. Steel safety slogan at that time was, "Knowing's Not Enough." It is not enough to *know* the rules for working safely. They must be *applied:* actively and correctly and fully. When an injury or fatality did occur in the plant, it was investigated quickly and thoroughly, not to fix blame, but to find out what had happened and to determine what changes in environment or work rules were needed to make sure that the same thing would not occur again.

Historically, we in the fire service have been quick to bury our dead with honors, but we have been very slow to investigate carefully the circumstances which caused the death to occur. Admittedly, industry can control the work environment while we must enter environments we neither control

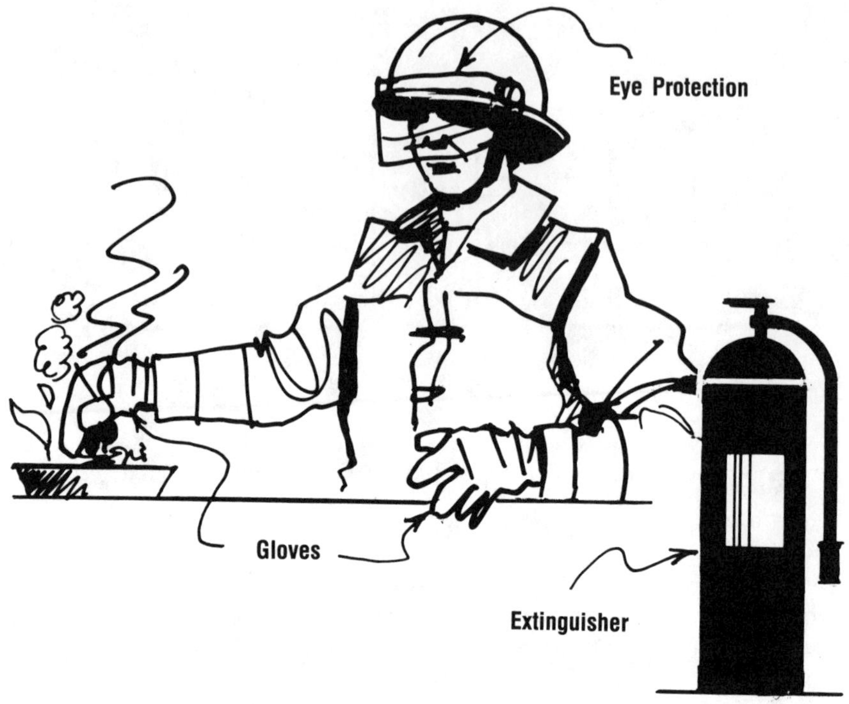

Fig. 2-14. Classroom safety.

nor, in many cases, have ever seen before. But *that* knowledge should make us *more* interested in safety, not less.

We need first, then, to examine our own attitudes toward safety and to commit ourselves as instructors to make a firm and complete break with fatalistic tradition: to make *our* classes and *our* students as safe as we can possibly make them. We will talk about changing student attitudes later on, and we will suggest that the role model *you* establish will play a big part in the attitude your students develop. It is so with safety.

Safety in the Classroom

It is a little difficult, truthfully, to get excited about safety in the classroom. Comparatively speaking, a classroom is a fairly safe place. We mentioned tripping hazards. Hot projector bulbs and hot projector parts will definitely cause burns, and we know of at least one case in which a projector bulb exploded and blew glass into the eyes of an instructor. Any demonstrations involving live fire in the classroom require eye protection, gloves, and an appropriate fire extinguisher (Fig. 2-14). Use proper techniques in lifting heavy equipment. Ground your electrical equipment, and use only electrical devices that are in good condition and approved for your purpose.

Safety on the Drill Ground

When we move to the drill ground, the environment becomes hostile in a hurry. Good fire training simulates the real world, and the real world is dangerous. This book is not a safety manual, and we do not have space to be exhaustive; use your common sense. Inspection and maintenance of ladders, breathing apparatus, rope, hose, and all other tools and paraphernalia is just as important for training as it is in emergency firefighting. Firefighters have been killed using gear that was retired from first-line firefighting but assigned to the academy because it was, ". . . good enough for training." That attitude is unforgivable. Don't settle for anything but the best.

Working with live fire requires a skillful and alert instructional team. Instructors must be positioned where they can take instant action if a key student, the person on the nozzle, for example, "freezes" or takes off. If you are working inside a purpose-built fire training building or at an actual building burn, maintain a constant, accurate head count. Know where *every* individual is at all times. There is no way to describe the panic that you feel when your training fire "blows up" and you suddenly realize that one of your students is unaccounted for. Don't let that happen. In structural training, fires should be started only by a designated instructor, and only approved fuels should be used. If accelerants are authorized, they *must* be materials with a high flash point. *Gasoline must be absolutely forbidden.*

In working with pipe-fed, pressurized flammable liquid or flammable gas fires, have an absolutely dependable and alert student or instructor handle the safety shutoff. This safety person must have a clear view of the entire evolution and complete authority to shut down the flow of fuel at the slightest indication that the procedure is not going as planned.

When working at heights, safety nets, air bags, or belaying lines should be used whenever possible. Most ladder work must be done without safety devices, so extra care and properly monitored student progress must be maintained. Ladder work and hose work, incidentally, seem to invite horseplay. There is nothing wrong with training being fun, but horseplay or a "lodge initiation" atmosphere, in which new recruits are harassed or pumped full of horror stories, have no place in a well-run training program.

The safety officer concept, delegating complete responsibility and authority for safety to one individual, should be applied in training whenever possible (Fig. 2-15). The safety officer should have enough experience to

Fig. 2-15. Drill ground safety.

recognize a hazardous situation quickly *before* it develops fully and must have the authority to stop instantly any training evolution in question.

Injury/Fatality Correlations

Monitor your experience in minor injuries. Don't let them go untreated or unreported. Industrial records prove that there is a direct correlation between minor injuries, major injuries, and fatalities. Have enough minor injuries, and a major injury is going to occur. Have enough major injuries, and eventually there will be a fatality (Fig. 2-16). Curbing minor injuries in training won't necessarily prevent a fatality, but concern over minor injuries will foster a positive attitude toward safety and help construct an environment in which a fatality is far less likely to occur.

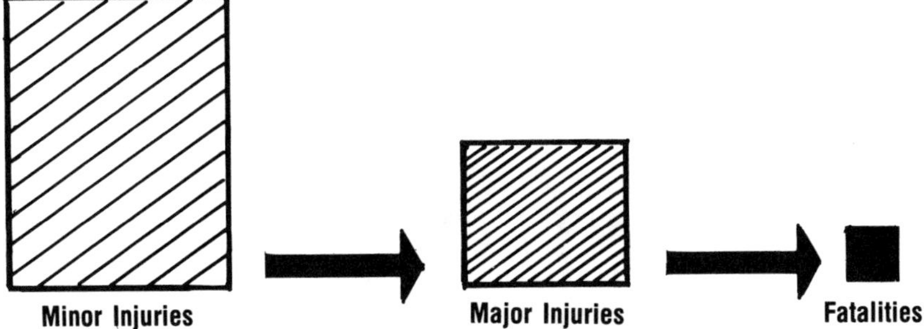

Fig. 2-16. Injury/fatality correlations.

After Working for the Best, Plan for the Worst

After doing everything possible to prevent a serious accident, plan what you must do and will do if and when one does occur (Fig. 2-17). Develop a simple but comprehensive standard operating procedure (SOP) to ensure that the best possible medical response will be employed, all necessary notifications will be made, statements of witnesses obtained, reports prepared, and similar essential actions taken. Establish, promulgate, and follow a firm policy that the next regularly scheduled training session *will* be held as planned. We hope and pray it will never be necessary, but if you ever face the need to follow these recommendations, pre-established plans and policies will steady you, and they will help pull you and your training program through what is always a terribly difficult time.

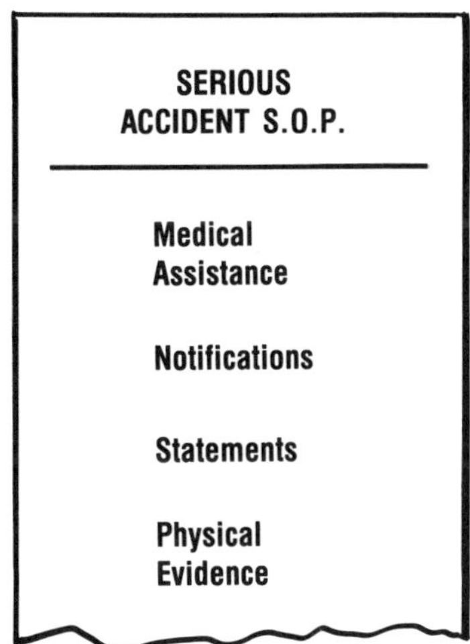

Fig. 2-17. Have a plan.

Fig. 2-18. Set the example.

Come On Now, Set a Good Example

Finally, *you*, that's right, *you*, turn in that burned and battered old helmet and that ragged turnout coat. Draw some new gear and keep it clean and in good condition. We all know you're a hard-nosed, head-down smoke eater; you don't need to advertise. Play it smart, play it safe, and set a good example (Fig. 2-18). Help *your* students to be alive tomorrow, next year, and the year after that.

So far we've looked at instructors and students and the environment in which we work. Before we tackle courses and lessons and methods of instruction, maybe we should take a look at how people learn. We're certainly not expert in educational psychology but, then again, not too many educational psychologists have ever humped a hose line up a smoke-filled stairway or have sat in the right-hand front seat of a piece of apparatus responding to an emergency. We've done both. Come on inside, and we'll share with you what we've picked up along the way.

3

Concepts of Learning

In all honesty, we approach this chapter feeling somewhat like a pair of brown shoes in a room full of tuxedos. So many scholarly books, so many university courses have been devoted to learning that it seems the height of presumption for a tired old firefighter to address the subject. As a matter of fact, let's not "address" the subject. Come into the kitchen, and we'll get some coffee and just sit and talk about it—maybe the bells will hit or the siren will blow and get us off the hook.

Teaching-Learning: The Importance of Change

You can consider the business of fire service instruction from either side of the lectern. From our side, the instructor's side, the process is called "teaching." In many respects it is like any leadership activity: guiding and encouraging others to follow some plan of action that we have decided is appropriate and correct.

From the other side of the lectern, the student's side, the process is called "learning." (See Fig. 3-1.) For learning to have any value, change must take place. If, at the end of a lesson, the student does not think or act or believe differently, if some change has not occurred, learning has not oc-

Fig. 3-1. Teaching-learning.

curred. Since we cannot *force* people to think or do or feel differently, there must be, or we must cultivate, in our students a *desire* to change, a desire to learn.

Several frequently quoted "laws of learning," originally developed by E.L. Thorndike, will give us some insight into the process by which people learn. No human behavior is absolute, of course, so these laws don't truly govern anything, but they help us to understand why and how people react to instruction.

The Law of Readiness

The law of readiness states that students learn best when they are *ready* to learn. As we said before, we cannot force people to learn. Since firefighters are usually strong-willed people, it would probably be more accurate to say that *our* students will learn *only* when they are ready to learn. Our first job is to motivate, to encourage our students to want to learn. Fortunately for us, in our particular work, this process is not too difficult.

Our big advantage as fire service instructors is that most firefighters are already motivated—highly motivated. We can reinforce their motivation, or perhaps help enhance motivation for unpopular subjects, by following a few simple steps.

1. Provide students with clear reasons, "why." Why is this subject in the training program? Why does the department or the

Fig. 3-2. "Because the Chief *says* so!"

training agency feel this material is important? Why is it scheduled at this particular time?

NOTE: The traditional fire department shout "Because the Chief says so!" should be used *only* as a last resort. It is considered customary to punctuate this phrase by hitting the lectern violently with either fist (Fig. 3-2).

2. Provide practical applications. Precisely and specifically, how will students use this information? How will this material make their work better or safer or easier? How does this information improve their chances for advancement or promotion?

3. Provide known objectives. Students should be told exactly what they are expected to do, the conditions under which they will be expected to perform, and what level or degree of performance is acceptable. Again, more later.

The Law of Effect

The law of effect states that people are drawn to those things which are, or appear to be, satisfying or rewarding. People return to experiences which are pleasing and avoid those which are unpleasant. The effect on learning is crystal clear and vitally important. Hugs feel good, and hot stoves hurt (Fig. 3-3). Both lessons are learned early, and both last a lifetime.

The need for affection and the need to be protected from harm are just

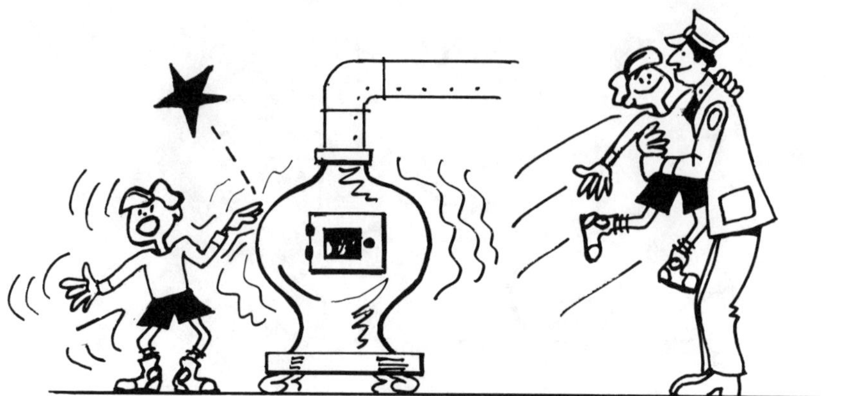

Fig. 3-3. Hot stoves hurt. Hugs feel good.

two of a complex range of needs which exist in every human being. Some of these needs are highly visible and self-evident. Others are more subtle, possibly even subconscious, but nonetheless important. The different degree to which each of us is driven by our own personal combination of needs determines how we react to challenges, to people, and to life in general. These same needs determine the way in which *your* students will react to learning situations.

Over thirty years ago, Abraham H. Maslow identified a "hierarchy" or ranking of human needs and theorized that these needs must be satisfied in an ascending order.[1]

    Self-actualization
    Approval—Esteem
    Affection—Belonging
    Safety—Security
    Physiological—Survival

According to Maslow, individuals generally will not function in, or devote attention to, higher needs levels until the needs below are at least partially satisfied. While most of the needs of our students must be fulfilled outside the classroom, we can put Maslow's theory and the law of effect to work for us in some simple ways during teaching-learning situations.

[1] A.H. Maslow: *Motivation and Personality*, New York, Harper and Row, Publishers, Inc., 1954

Fig. 3-4. Break time.

Physiological—Survival Needs

The basic human needs for food, drink, shelter, and rest can easily be met. Periodic breaks, ten minutes at the end of fifty minutes of teaching, or any similar reasonable and regular system, will assist students and the instructor to stay alert and interested (Fig. 3-4). Breaks give students an opportunity to relax, to get a drink or a snack, and to use the rest room. Instructors need these same opportunities, plus a few minutes to check the lesson plan, to determine if they are ahead of or behind schedule, and, quite often, to set up visual aid equipment for the next session. Therefore, be selfish with *your* break time. Do not be reluctant to tell students, in a courteous but firm way, to hold any questions or discussion until the class reconvenes.

Reasonable student comfort should be maintained, as we suggested in Chapter 2. Specific to Maslow, temperature extremes should be avoided, aside, of course, from live fire training. We have seen firefighters cooped up in un-air-conditioned classrooms in August with sweat dripping from their chins, and we have seen firefighters shaking with cold during lectures in unheated drill ground buildings in January. Neither practice makes sense. Students intent on merely surviving are not going to learn very much. If it is advisable or necessary to drill during extreme weather conditions, use the time in the field to practice known skills. Teach new material in a comfortable environment which will permit and encourage students to focus on you and your instruction, rather than on their own misery.

Safety—Security Needs

Students need to feel safe, confident in the equipment and confident in your ability. If your students know you blew a hole in the classroom ceiling the last time you taught the line gun (Fig. 3-5), they will probably be more

Fig. 3-5. "If you blew a hole in the ceiling . . . "

interested in staying out of the line of fire than in any teaching points you are trying to make.

Students also need to feel secure in their own ability, to be confident that they will succeed. Contrast, in your own mind, the apprehension of the novice to the confidence of the experienced firefighter: working at height, working in breathing apparatus, or working with live fire. All are aspects of training in which poor handling by an instructor can drive a student literally to the "point of no return." Think about it. The person *you* run out of the drill yard might have the potential to become an outstanding firefighter, given a reasonable opportunity to overcome natural fears.

Build in opportunity for early success and frequent success. Confidence and familiarity control fear. Give your students the gradual learning opportunities they need to gain that confidence and familiarity. Not many of us would have made it up the aerial ladder if we hadn't started on the 24-foot extension.

Affection—Belonging

The fire service is a unique and special fraternity. Capitalize on it. Identification with the department or a battalion, a company, or a specific recruit class can be a powerful motivator. The ongoing competition and needling between the engine companies and truck companies is all a part of

Fig. 3-6. "Belonging!"

our human need to identify with something larger than ourselves, to "belong." If you have any doubt, just start counting shoulder patches (Fig. 3-6). And remember, the people who ride the truck are bigger and smarter and do all of the *real* work.

Approval—Esteem

The human need for recognition, for commendation, for praise gives us one of our best ways to put the law of effect to use. Everybody likes a pat on the back. We used to call them "Attaboys," an old-fashioned term for any words of approval or praise from the instructor. Each of those opportunities for early and frequent success we build into the program gives us a chance to say, "Good," or "You've got it," or "That's it, nice work!" Your approval, your encouragement will move students farther and faster than any other action you can take, and it doesn't cost a penny.

One note of caution—for praise to have any worth, it must be merited. Don't praise poor performance. Give the student having difficulty your encouragement and your assistance, but save your words of approval for when the student gets the work right.

Fig. 3-7. "Making it" feels great!

Self-Actualization

At the top of the list are those needs which drive us to do and be our very best. The greatest thrill in teaching is in seeing your students become truly proficient, particularly if you have played a strong role in helping them to achieve beyond their own expectations. Our society consistently rewards mediocrity. The accepted standard, all too often, is to be "just good enough to get by." Superior instructors should set their sights much higher.

Strive to instill in your students a sense of craftsmanship. Firefighting is as much an art as it is a science. Knowing how to do a job is only the beginning of learning. Knowing how to do that job perfectly, under any and all conditions, with an absolute minimum expenditure of energy and effort, is the mark of a craftsman (Fig. 3-7). Our goal should be to develop in our students the desire to extend themselves mentally and physically just as far as they can. Maslow assures us that our students have the *need* to achieve. Our job, as instructors, is to assist them to see the highest potential they possess and to help them to reach that potential successfully. The rewards will be twofold. The fulfillment you assist your *students* to find will provide *you* with your own "self-actualization," the opportunity to reach the personal goals you establish for *yourself* as a fire service instructor.

We seem to have spent a lot of time on the law of effect. People return to those experiences which are pleasing and avoid those which are unpleasant. The concept is important and if we apply it well, everybody wins—the students, the instructor, the department, and, ultimately, the public. That's a parlay worth shooting for!

The Law of Primacy

The law of primacy states that the first experience in any activity is of paramount importance. As you introduce new topics or new skills in your classes, take special care to avoid any error in fact or sequence or manipulation. Ensure that this first exposure of your students to any new area of learning will be totally complete and correct. First impressions, as always, are the most important.

When students begin to apply new information, arrange early exercises at relatively slow, relatively easy, achievable levels. Make sure that your students have every possible opportunity to perform correctly on their first attempt. It reinforces learning, promotes confidence, and provides that early success which is so very important.

The Law of Intensity

Worth a brief mention is the law of intensity, which states that people remember best those things which are most vivid. Each of us has memories which are exceptionally graphic because the circumstances surrounding the remembered event were particularly intense—particularly frightening, particularly painful, or particularly joyful. In order to capitalize on this law, we need to build intensity into our lessons and aids and simulations.

Reference to specific "real-world" emergency situations provides a vivid bridge which will assist your students to make the transition from abstract concepts to practical applications. Colorful training aids with lots of "pizzazz" reinforce retention. Exercises and simulations of all types should be as real as you can make them, within the bounds of safety, common sense, and reasonable cost (Fig. 3-8).

One note of caution: we are *not* advocating "learn or burn" training. In the past, in some training programs, structural firefighting instructors held competitions to see who could build the biggest, hottest training fires. Students, caught in the middle of this game, were often pushed beyond the limits of endurance which could reasonably be expected of novice firefighters. There is absolutely no doubt that the law of intensity came into play, but the net effect was negative rather than positive. Students thrown into these situations without adequate experience often went away con-

Fig. 3-8. The law of intensity works on the drill ground.

vinced that interior firefighting was absolute insanity, rather than an effective method of fire attack. Make your training program "like real," but also *make it appropriate* for the proficiency level of your students, and make it safe for all concerned.

The Law of Exercise

The final law of learning we will discuss is the law of exercise. Stated simply, it reminds us that the things most often repeated are the things best remembered. Retention improves with practice, practice, and more practice—not just walking through the same process, over and over again, blindly and without thinking, but participating in *accurate, motivated* practice. Simple repetition won't do anyone any harm, but repetition by students intent on improving their skills, motivated and encouraged and goaded (carefully), by the instructor, will accomplish much more. As Vince Lombardi put it: "Practice does not make perfect; perfect practice makes perfect" (Fig. 3-9). And practice makes permanent.

Fig. 3-9. Practice, practice, practice.

Learning Plateaus

Learning does not increase on a regular, smooth curve. Individuals and classes will often make a marked upward swing, rapidly advancing in new skills or concepts, and then come to a point where the learning curve flattens out or enters a "learning plateau." When this stage occurs, additional teaching or continued practice may be of little or no benefit. A plateau may be caused by fatigue or perhaps by a perception that the training material has taken an uninspiring turn (Fig. 3-10). The important thing to remember is that "flat spots" *do* happen and to be alert for them in your classes. If you sense that student interest or learning has flattened out, take a break or change the activity or subject to something different. Another alternative is to remotivate your students, to give them a motivational booster shot. There is no general solution. The key issue is to recognize a learning plateau when it occurs and to "shift gears" rather than to let the whole teaching-learning process grind to a halt. Your students will benefit and it will make your job a lot easier.

Type of Information Which We Teach

As fire service instructors, we are engaged in a continuing and dynamic process of transferring information from our heads and hands to the heads and hands of our students. Along with other educators in many different disciplines, we recognize three fundamental types of information which we transfer, or teach: knowledge, skills, and attitudes (Fig. 3-11).

Knowledge

Knowledge, simplified, is being aware of and being able to recall something, to understand and apply information. You will transfer to your

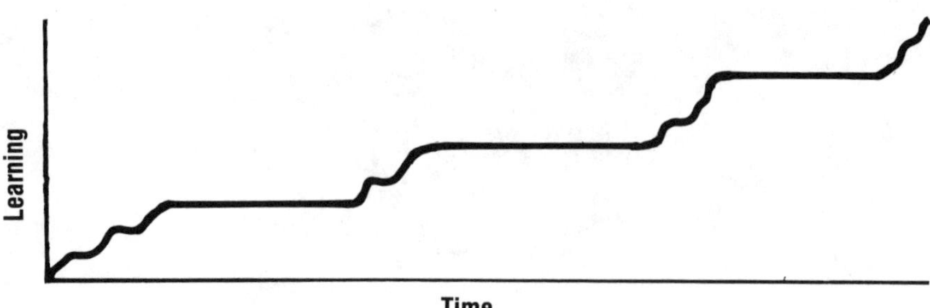

Fig. 3-10. Learning plateaus.

students knowledge of facts: water weighs 8.35 pounds per gallon; knowledge of persons: John Jones is the state fire marshal; knowledge of things: this is a halligan tool. You will teach concepts: this type of motion illustrates centrifugal force and this is how centrifugal force is used in the movement of water. You will teach principles: this is what is indicated when the windows of a fire building are blackened and cracked and the smoke is thick, oily-looking, and yellowish-brown in color. In addition, you will teach a wide range of intellectual activities: for example, creating in your students the ability to solve hydraulics problems, or to construct a pre-fire plan, or to analyze a fireground situation and apply the appropriate rules of tactics and strategy.

The acquisition of knowledge is a lifelong task. Your students will obtain knowledge from you in formal class sessions. They will also obtain

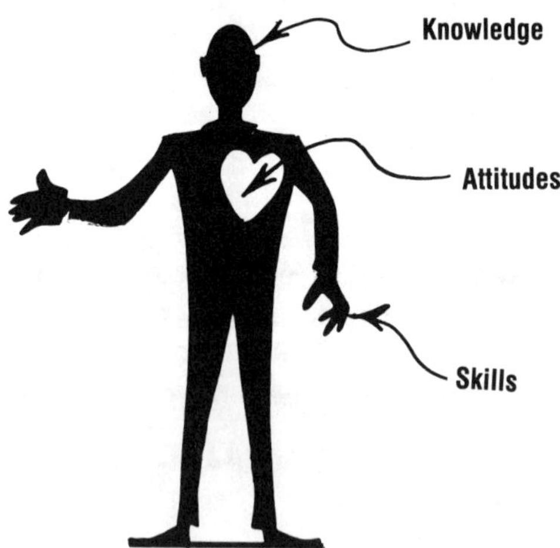

Fig. 3-11. Head, hands, heart.

knowledge in informal settings—over coffee or during breaks. Finally, knowledge is obtained through experience, an effective but often very expensive way to learn.

Skills

A skill is a developed ability to perform a particular manual task or physical act, to "do" something. You will teach your students to use, to manipulate, to adjust, to operate—to carry out an infinite variety of physical activities. Our goal is to create in our students the ability to perform both physical and mental skills: with *ease*, comfortably and swiftly; with *precision*, carrying out all steps in the proper sequence; and with *accuracy*, correctly producing the desired results.

Attitudes

Attitudes are quite different from knowledge and skills and cannot be "taught," in the same sense that knowledge and skills are taught. Attitudes are concerned with beliefs, ideals, values, preferences, and similar abstract human characteristics. As knowledge and skills reside in the head and hands, attitudes may be said to live in our hearts. Our beliefs may not be easily defended on logical grounds, but they govern much of what we are willing to do and the manner in which we do it.

We may want our students to accept the idea of building on worthwhile traditions, or the idea that safety must temper aggressiveness on the fireground, or the idea that service to the community should be paramount. Dogmatic statements by an instructor will not, in all probability, cause adult students eagerly to embrace such abstract concepts, particularly if the ideas proposed seem contrary to their prior personal experience. To change ideals and attitudes, we must give firefighters an opportunity to discuss, probably to argue, the pros and cons. Even with adequate discussion, the process of attitude alteration may be long, slow, and laborious. New knowledge and new skills are accepted by all of us much more readily than we accept new ways of thinking or different views of familiar subjects. Perhaps the best teaching tool in these subtle areas is personal example. *Your* attitudes, your beliefs, your convictions, the role model you establish will go far in establishing the attitudes and beliefs of your students.

The Senses: Channels for Learning

People can learn or acquire knowledge, skills, and attitudes only

Fig. 3-12. Five senses.

through the five human senses: by seeing, hearing, touching, tasting, or smelling (Fig. 3-12). We have read a number of publications which assign specific percentages of learning to each of the senses. While we don't believe it is possible to state, flatly, that people acquire 85 percent of what they learn through sight, for example, we *do* agree that most learning, unquestionably, comes through a combination of sight and hearing. Touch, of course, plays a large part in transferring manual abilities. Smell and taste are not used very often in fire service instruction but may be necessary in specialized areas such as chemistry or arson investigation.

Seeing versus Perceiving

There is an important distinction between seeing and perceiving and between hearing and understanding. We learn through seeing, true, but seeing alone may not cause us to *perceive*, to take hold of, to attain awareness and understanding. Many of us walk up and down stairs daily, perhaps several times daily, at home or at work. Yet few of us know how many stair steps make up that stairway. We *see* the steps, but we have no reason to *perceive* how many are there. We are willing to bet that the next time you climb that set of steps you will count the number; you now have a reason to perceive, a motivation, even if it is no more than idle curiosity.

This concept is far from new. "Hear ye indeed, but understand not; and see ye indeed, but perceive not." So wrote the prophet Isaiah, nearly three thousand years ago. As fire service instructors, we must not only provide our students with an opportunity to see and hear; we must also ensure that they perceive and understand the knowledge, skills, and attitudes we are presenting to them.

Kinesthesia

Another important concept concerns kinesthesia: the perception of one's own muscular movements. In teaching physical skills we can show students how to hold a particular tool—how to position their hands and how to take the correct stance. What we *cannot* readily show them is the amount of force they must use or the muscular coordination required to move the tool properly. Using a fire axe effectively is a good example. Chopping without bringing the axe back overhead forces the firefighter to *punch* the axe from head height into a roof or floor. This limited motion requires that the axe be pushed down, rather than pulled down from a full overhead swing. The shoulder provides most of the force rather than the arms—not an easy concept to convey. You can demonstrate the action, but there is no convenient and practical way to "teach" the kinetics involved. Trying to transfer such a perception of muscular movement can be extremely frustrating to both the instructor and the student. Fortunately, learners seem to sense intuitively when they achieve the correct motion. It *feels* right, and, when it feels right, it probably *is* right.

Organization: A Critical Key to Learning

So far in this chapter we have discussed a number of laws of learning, the types of information we teach, and some of the mechanisms by which we transfer information to our students. Let's consider how the information must be organized to create the best opportunity for our students to learn easily and effectively.

In any fire service topic or subject area there are dozens or hundreds or, perhaps, thousands of bits of information which must be conveyed. Unorganized, they have little or no meaning and could be learned only as single, seemingly unrelated elements—a difficult task at best. John Hoglund, director of the Maryland Fire and Rescue Institute, once compared this mass of information to a basketful of fragments of colored glass. Unorganized, the fragments have no meaning. Properly organized by a skillful craftsman, they can be arranged to form a stained glass window which carries a message for every person who stands before it (Fig. 3-13). So it is with teaching. *You* are the artist, organizing, arranging, tying together, relating one piece of information to another to make a coherent and understandable whole.

Your teaching points and the flow of lesson information should be arranged in a logical and clearly understandable sequence. Students must be able to recognize immediately the relationship of each item to the points which precede and the points which follow as the lesson unfolds, step by step.

Among the basic concepts of material organization are: known to unknown, simple to complex, and whole-parts-whole.

Fig. 3-13. Arrangement of the pieces is everything. *Dede Bachtler; courtesy Maryland Fire and Rescue Institute.*

Fig. 3-14. Known to unknown.

Known to Unknown

One organizational technique is to start with a base of known information. Make sure that the students have their feet planted, figuratively, on familiar ground. Previously learned lessons, standard operating procedures, and common knowledge are all acceptable "known" foundations. From this solid base you can push off, step by step, point by point, into new and unknown areas (Fig. 3-14).

Simple to Complex

This pattern of organization moves from simple, easily understood points to levels which are increasingly complex and difficult. If the transitions are carefully and effectively arranged, with built-in periodic checks to ensure understanding, students are led, step by step, into the more complicated lessons or lesson segments without becoming confused or lost (Fig. 3-15). Interior structural firefighting is an example which we will use in a later chapter. Before we can teach interior structural firefighting, students must first master a lengthy chain of fundamental skills.

How fast the process moves depends on the complexity of the material being taught and on the capability of the student group. These two factors

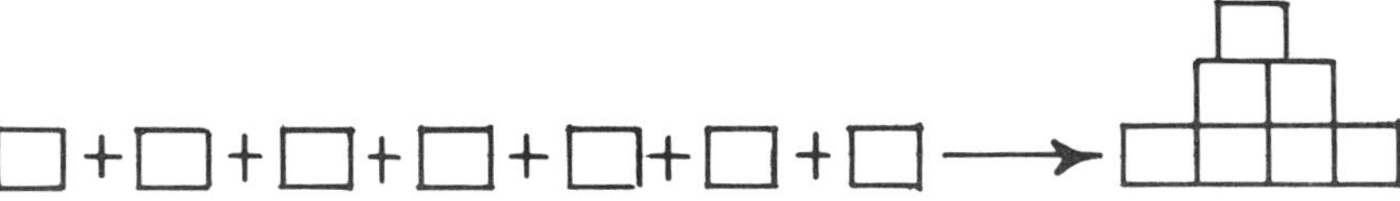

Fig. 3-15. Simple to complex.

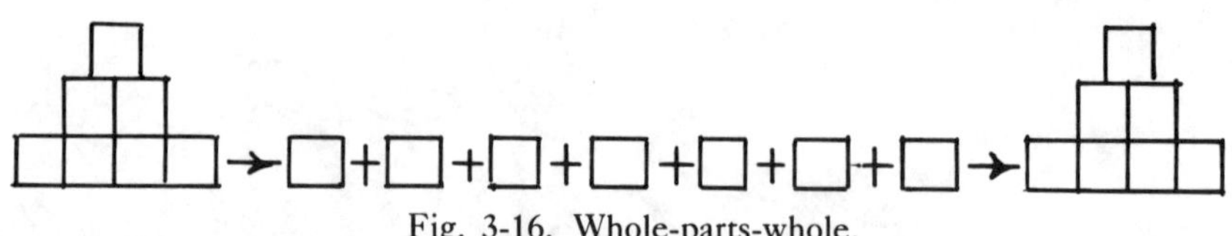
Fig. 3-16. Whole-parts-whole.

will establish both the number of steps necessary and the amount of time required.

Whole-Parts-Whole

As the title suggests, students are first shown the whole operation or concept. In technical or classroom subjects, the "whole" can be displayed through lesson objectives or by an overview or synopsis of the material you intend to cover. In skills instruction, the "whole" is displayed by first demonstrating the entire operation at normal speed. Reduced speed demonstrations are then performed, in which the "parts," the main steps and key points, are identified. Student practice follows and, when the individual parts have been mastered, they are joined together to form the "whole" skill or task (Fig. 3-16).

Whole-parts-whole provides something to imitate, step-by-step instruction and practice in the individual parts, and, finally, an opportunity to combine it all into a useful operational sequence. We will go into more detail in Chapter 6.

Step by Step, Point by Point

We believe that *logical organization of teaching points is one of the most important aspects in helping students to learn.* Technical experts who do not know how to arrange their material in a logical and understandable sequence often fail at teaching. A good instructor, even though less technically expert, can succeed with the same material by organizing it effectively. Clearly identified teaching points, logical sequence, distinct transitions from one topic to another, rational grouping of subjects and skills all lead to superior, step-by-step instruction which will provide your students with a valid and profitable learning experience.

The View Across the Lectern

At the beginning of this chapter we suggested that it might be useful to

Fig. 3-17. "Making ready" to learn.

view fire service instruction from both sides of the lectern: the instructor's side and the student's side. Before we leave the subject of learning, let's return to that thought for just a moment. While the instructor and the student report to the same location at the same time to begin the same class, there is a considerable difference in perspective or viewpoint, depending on whether you are standing behind the lectern or sitting in front of it.

We, the instructors, arrive:

• concentrating on the class ahead; our thoughts and attention tightly focused on the material to be covered.

• enthusiastic, motivated, eager to begin

• confident in our own ability and secure in the knowledge of what we have planned.

The students arrive:

• each thinking of something different with little, if any, specific focus on the class which is about to begin

• probably not too enthusiastic—perhaps, perhaps not, depending on whether they were detailed to be there or have come because they *want* to be there—in either case, less enthusiastic than the instructor

• lacking in security—not knowing what they will be expected to do nor how well they will be expected to perform

From the security of our side of the lectern, it is our responsibility to see that the attention of our students is focused on the importance of the subject material and that they are, to the best of our ability, made *ready* to learn (Fig. 3-17). The steps we suggested at the beginning of this chapter to create, enhance, or reinforce motivation were: *(1)* to provide clear reasons "why," *(2)* to provide practical applications, and *(3)* to provide known objectives.

Leave your coffee cup over in the sink. We'll go back to the office and take a good look at how those three elements, in particular, and the concepts of learning, in general, tie into the course development process.

4

Course Development

Every city, county, and state fire training program has a series of "bread-and-butter" training courses: fundamental subject areas common to all progressive fire service agencies. Among these courses are usually such standard topics as:

> Basic Firefighter Training
> Advanced Firefighter Training
> Officer Training:
> > Tactics and Strategy
> > Leadership Skills
> > Management and Administration
> Specialist Training:
> > Apparatus Operators
> > EMS Personnel
> > Rescue Personnel
> > Public Educators
> > Inspectors and Investigators

As fire instructors, we are often called upon to rewrite courses similar to these or to create new programs in response to the changing demands on our departments or training agencies.

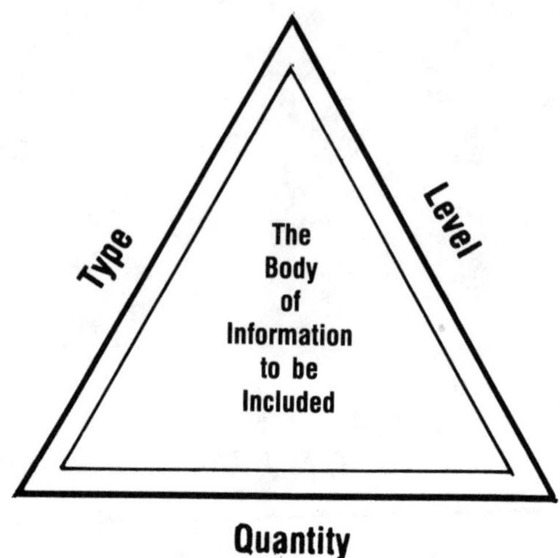

Fig. 4-1. Course development considerations.

Development of a new training program in a small department, with limited resources and limited personnel, may be as simple as sitting down with a pencil and a legal pad and roughing out an instructional strategy. A lot of excellent teaching has started in just that way. In a large department or agency, with complicated needs and a well-organized and well-staffed training structure, course development may be a much more complex process.

Types, Levels, and Quantities of Information

We will outline the major course development steps which should be considered (Fig. 4-1). You or your jurisdiction may choose to simplify the process, or you may find it necessary to add additional steps and check points, as your particular needs require. Before we can construct a course development process, however, we must establish some foundations. In the last chapter, we introduced three differing types of instructional information. We will expand a bit on that concept and also talk about differing levels of information and differing quantities of information. Let's consider some of the concepts and some of the terminology involved.

Types of Information

In Chapter 3 we discussed the idea that the instructional information we

Knowledge = Cognitive Domain

Skills = Psychomotor Domain

Attitudes = Affective Domain

Fig. 4-2. Domains of learning.

teach can be classified as knowledge, skills, or attitudes. In formal educational settings, these three categories are referred to as the "domains of learning," and they are designated as cognitive, psychomotor, and affective, respectively. The cognitive domain involves objectives in which the student is required to recall or recognize knowledge, to demonstrate mental or intellectual skills. The psychomotor domain involves objectives in which the student must demonstrate the ability to perform a manual or physical skill. The affective domain involves objectives in which the student must relate to or describe attitudes, interests, values, or appreciations (Fig. 4-2).

The terminology you prefer to use locally is entirely up to you. We will use knowledge, skills, and attitudes for most of our work together, but a recognition of the terms cognitive, psychomotor, and affective is useful, particularly in exploring the idea that there is, within each of the domains, a hierarchy or a series of levels of information, ranging from the rather simple to the highly complex.

Levels of Information

Many fire service education and training agencies are currently identifying the educational levels of their lesson material based on a system developed by a consortium of educators in the early 1950s. Working under the editorship of Benjamin S. Bloom, this group published a handbook, establishing and explaining the concept that we can identify not only the *type* of information we are planning to teach, but the *level* of complexity as well.[1]

[1] Benjamin S. Bloom, Editor: *Taxonomy of Educational Objectives*, New York, David McKay Company, Inc., 1956

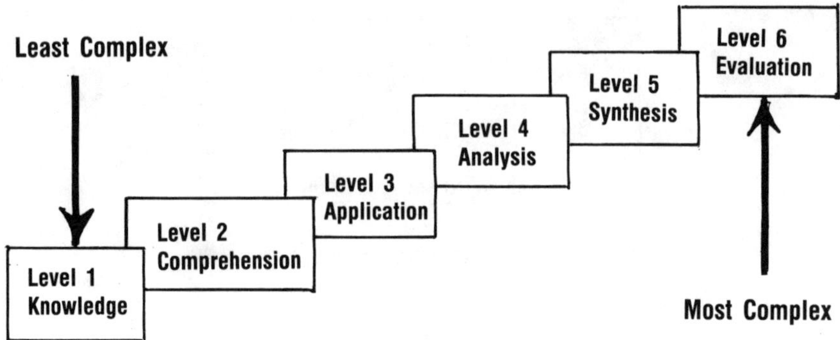

Fig. 4-3. Bloom's cognitive taxonomy.

Bloom and his colleagues worked almost exclusively in the cognitive and affective areas. For us, the most useful aspect of their work concerns the cognitive domain (Fig. 4-3). They proposed six levels, with one being the lowest or least complex and six the highest or most complex. These levels are outlined in the paragraphs following.

Even in simplified form, these ideas are kind of tough going, like trying to get at a smoky fire on the second floor. Hang in there—we're right behind you.

Level 1: Knowledge

Knowledge objectives emphasize the psychological processes of *remembering*. Knowledge involves the ability to recall specifics and abstractions.

Knowledge of Specifics:
- Knowledge of terminology
- Knowledge of facts
- Knowledge of rules and practices
- Knowledge of methods and procedures
- Knowledge of trends and sequences
- Knowledge of classifications and categories
- Knowledge of criteria

Knowledge of Abstractions:
- Knowledge of principles and generalizations
- Knowledge of theories and structures

Level 2: Comprehension

Comprehension objectives involve the lowest level of *understanding*. Comprehension refers to ability to understand what is being communicated and ability to apply the information or idea received. Comprehension may be demonstrated by:

- Translation: The ability to restate, paraphrase, or change the form of information or ideas
- Interpretation: The ability to explain or summarize information or ideas
- Extrapolation: The ability to extend or project data or sequences on beyond the information received

Level 3: Application

Application may be described as the ability to remember and *to use* methods, procedures, rules, ideas, technical principles, or theories.

Level 4: Analysis

Analysis is concerned with the ability *to break down* a concept into its component parts, so the ideas or elements contained are made clear and/or the relationships between the ideas are precisely described.

- Analysis of Elements: Identification of the ideas or elements included
- Analysis of Relationships: Identification of the connections and interactions between elements and parts

Level 5: Synthesis

Synthesis involves *putting together* elements or parts to form a whole. It includes the process of working with pieces, parts, and elements, arranging and combining them to form a pattern or a structure not clearly present before.

The ability to synthesize may be demonstrated by:

- Development of a communication in which the writer or speaker assembles and conveys ideas, feelings, or experiences to others

• Development of a plan or work or a plan of operations
• Development of a set of relationships which classify or explain particular data or phenomena

Level 6: Evaluation

Evaluation concerns *judgments* about the value of material and methods for given purposes, including quantitive and qualitative judgments about the extent to which material and methods satisfy criteria.

• Judgments in Terms of Internal Evidence: Evaluation from such evidence as logical accuracy, consistency, and other internal criteria
• Judgments in Terms of External Criteria: Evaluation with reference to selected or remembered criteria

Nobody Said It Was Going to Be Easy

We have just gone through a mass of relatively dense information, most of which is, probably, entirely new to you. Some of the wording may be somewhat unfamiliar and difficult, but the concepts, if you examine them closely, are really quite simple and they are extremely useful. If you go back and reread the summary carefully, you will see many places where fire service educational points fit clearly and comfortably within the system.

We felt it necessary to include enough detail to clarify thoroughly the meanings of the six levels. How much of the detail you use will be determined by your needs. There is nothing magic about even holding the number of levels at six; IFSTA, for example, in the fourth edition of *Fire Service Instructor,* mentions only five, eliminating evaluation.

It is certainly not necessary that we use all of the subdivisions derived from Bloom's taxonomies, nor, in all honesty, is it essential that we achieve 100 percent accuracy in fixing our lessons at any specific and precise level. Gaining experience with the concept will increase your ability to use it wisely, as well as making it more familiar and, perhaps, less threatening. In the meantime, who is going to argue with you if you get just a little bit off base in some subtle difference between Level 2 and Level 3? What *is* important is to recognize what "Level 2" or "Instructional Level 3" *means* when you see these designations, and to be able to employ what you need among these educational tools in order to develop courses and lessons which will best serve you, the other instructors in your system, and your students.

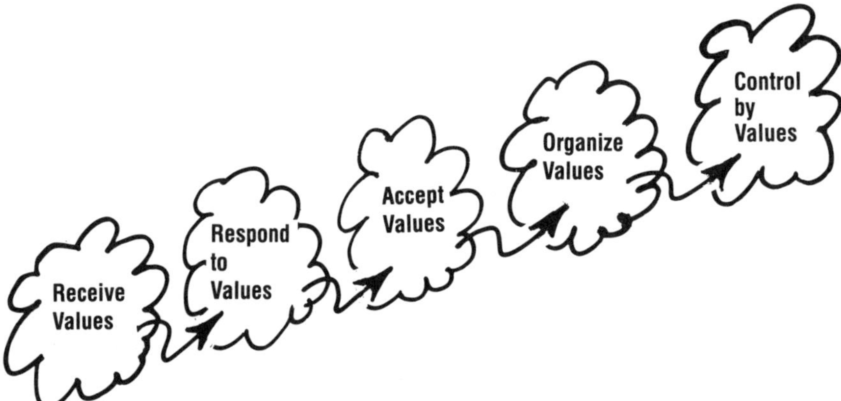

Fig. 4-4. Bloom's affective taxonomy.

The key point we want you to remember in developing courses, and in planning and teaching lessons, is that *there is a substantial range of difficulty or complexity in the information we present.* In course development, therefore, we must carefully examine and adjust the *amount* of material we include, the *pace* of presentation we plan, and the *method* of instruction we select, to best reconcile the *difficulty* of the instructional material with the *needs* and *abilities* of our students.

The Affective Domain

This area is significantly more complex and even more difficult to deal with than the cognitive. In Bloom's *Taxonomies*, the five levels within the affective domain are identified as: 1, receiving; 2, responding; 3, valuing; 4, organization; and 5, characterization by a value complex. *Greatly* simplified, levels one through three concern the willingness of an individual *(1)* to receive, *(2)* to respond to, and *(3)* to accept values. Level four deals with abilities to conceptualize values and to organize a personal value system. Level five concerns control of the behavior and philosophy of the individual by the value system accepted (Fig. 4-4).

If we look carefully, we can see that these levels describe the sequence of acceptance we must establish in changing the attitudes of our students. It is probably not necessary, though, that we dig deeply in this area. If you encounter a particularly difficult problem in creating attitude change in your students, you may want to work your own way through Bloom or some similar in-depth study. We will leave it at that.

The Psychomotor Domain

The area of skills instruction fills a high percentage of our time and interest in fire service training. Unfortunately, it is the educational area in which the least research has been accomplished. The few lists on psychomotor learning that we have seen have been more concerned with the patterns of *how* students learn rather than with *levels* of learning.

For example, at a 1974 University of Maryland Fire Instructors Seminar, then faculty members Dr. Stephen Wolk and Dr. Cyril Svoboda presented a series of five levels of action or instruction involved in teaching skills:

I: Imitation: Provide the learner with an opportunity first to make an internal mental rehearsal and then to duplicate an action as it is observed

II: Manipulation: Provide the learner with an opportunity to perform the action according to instruction, rather than on the basis of observation

III: Precision: Provide the learner with an opportunity to practice and develop accuracy and precision in performing the action

IV: Articulation: Provide the learner with an opportunity to practice and establish a sequence of action which flows smoothly

V: Naturalization: Provide the learner with an opportunity to practice and develop a high level of proficiency with the least wasted motion

The older we grow, the more we are inclined to believe that there is nothing new under the sun. In the summer of 1953, George Orgain taught a group of us at Oklahoma A & M a time-tested method known as "Show and Tell." "Show 'em how" (I, above), "Tell 'em how" (II, above), "Let 'em do it" (III, IV, and V, above)—nothing really changes.

In a way, the fact that educators are concerned with *how* students learn psychomotor skills rather than being concerned with *levels* of psychomotor learning isn't surprising. A skill, most certainly any emergency service skill, must be performed completely and correctly. Individuals may differ in the

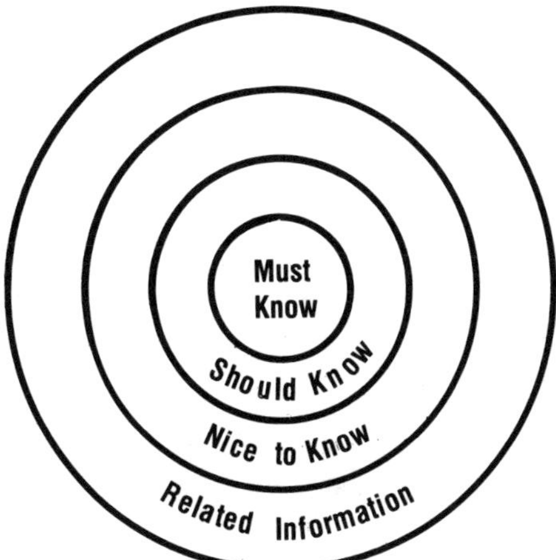

Fig. 4-5. Training target.

speed or the fluidity with which they perform a particular task, based on experience or personal physical characteristics, but, in the final analysis, they either *can* perform the skill or they *cannot*. Our teaching goal, therefore, must be to bring the student to an acceptable level of psychomotor or skill performance, not to some intermediate point which is "good enough."

With our interest in skills instruction, this is an area in which some future fire service instructors could make an excellent contribution to the field of education, by studying and publishing information on acceptable *levels* of skills proficiency if, in fact, these levels do exist and can be identified in some useful fashion.

Quantities of Information

Finally, in addition to types of information and levels of information, we need to consider the quantity of information included in a particular course or lesson. Picture a typical target, with the mass of information in this specific subject area divided among the rings (Fig. 4-5).

Those things which the student *must know* form the center of the target—the "bull's-eye"; the first ring contains information that the student *should know* in order to function effectively; the second ring holds those items which are *nice to know;* and the third ring is composed of *related information*. The area surrounding the target can represent the sea of information which may be important in our field but is unrelated to this particular course. Our

Fig. 4-6. Many things bear on course content.

task as course developers and teachers is to concentrate the "in-class" time and attention of our students at or near the center of the target.

Centrifugal versus Centripetal

In the fire service classroom, instructors and students often find themselves talking about the same subject, but moving in opposite directions. The thought patterns, questions, and discussion of students tend to be "centrifugal"; they move *away* from the center of the target. Our students, because they are keenly interested in *all* things concerning the fire service, will steer the discussion down all sorts of fascinating paths which are fun to explore but which take us away from the subject at hand. Instructors, therefore, must provide an even greater "centripetal force," focusing class time and attention on the central issues of the lesson. Strong, well-constructed, and logical course guides and lesson plans are tremendous assets in keeping things "on target."

How Much Is "Enough"?

The overall size of the target, the amount of information we can crowd into each ring, and how far out on the target we can afford to go are decisions determined by a number of considerations, some of them political, some of them involving the budget (Fig. 4-6).

First There Is the Issue of Time . . .

In developing courses for the volunteer fire service, a very real and legitimate political question concerns the amount of time prospective students are willing, or can reasonably be expected, to invest. With the proliferation of training opportunities in recent years, plus the increasing sophistication and complexity of the training offered, we are rapidly draining the available pool of volunteer training time. Adding to the problem is the reality that many of our volunteer people are cross-trained in emergency medical skills, confronting in that discipline another heavy volume of necessary, and in many cases required, training.

In both volunteer and career training programs, the issue of certification and the quantity of training necessary to achieve and maintain certification are also part of the political course development equation. It is easy, and it sounds professionally high-minded, to insist that educational goals stand above individual desires and organizational politics. Unfortunately, the process just does not work that way. The perceived needs and desires of our constituents and their organizations *will* be made known. It makes much more sense to seek these out in advance, rather than to discover them only after a course of instruction has been developed and released as a part of your curriculum.

. . . And Then There Is the Budget

Dollars determine a number of course development issues. How much can we spend on the course development process? How much can we spend on creating instructors' guides, student materials, and visual aids? More significant, in the long run, will be the continuing costs: money for reproduction of materials, for instructor teaching time, and, in career or industrial fire departments, for hours personnel spend in the classroom or on the drill ground.

The quantity of information we can put into the target *is* limited. In course development, we must strive to get what some folks call, "the biggest bang for the buck." We must concentrate class time in those areas which the student *must* know and *should* know. The "nice to know" and related information must often be pushed outside the classroom, through outside reading assignments, or they must be eliminated entirely.

One More Foundation Stone

Of the three foundation areas we have discussed, the last area, quantity of information as reflected in political and budgetary concerns, will influence the course development process directly. Political and budgetary review points will appear clearly in the course development checklist. The first two

Fig. 4-7. The foundation stone.

areas, types of information and levels of information, will be less visible because they affect the technical development of instructional material, specifically, course and lesson objectives. Since "objectives" form the backbone of the course development process, we must define the term and describe how objectives are constructed (Fig. 4–7). We promise this topic will be the *last* one we talk about before getting into course development. After all, that *was* the title we selected for this chapter.

Objectives

One of the most important concepts to come along during our years in fire service training is the identification and acceptance of the concept of performance objectives. Please don't become involved with whether the correct term is "behavioral" objectives, "performance" objectives, or any other similar phrase. Ever since we have been in this work, the terminology has been changing, and it will continue to change. We will use "performance objectives" because we feel that term most accurately describes what we are intending to convey. If your jurisdiction uses something else, that's fine.

We feel the best text on understanding and creating objectives is Robert F. Mager's excellent little book, *Preparing Instructional Objectives*.[1] If you have not read it, we urge you to get a copy and do so. It can be read in a short evening, and it is a lot of fun. This book had a tremendous effect on our

[1] Robert F. Mager: *Preparing Instructional Objectives*, second edition, Belmont, California, Fearon Publishers, Inc., 1975

teaching in that it forced us to think in terms of what the student receives, rather than in terms of what the instructor does.

Before reading Mager's book, we would think of a teaching assignment in terms of, "What will *I give* them?" "What folder will *I* pull from my file to *present* to this particular group of students?" Our thinking was *instructor*-centered rather than *student*-centered. In years past, instructors accomplished some reasonably effective teaching in that fashion, and we and the students usually came away feeling we had enjoyed the class. Many of us might have been hard pressed, however, to establish a lengthy list of specifics which were actually *learned*.

Writing Performance Objectives

Performance objectives force you to consider, very specifically and very accurately, what you want the student to *learn*. As we plan a course, as we prepare a lesson, as we begin to teach, we must be ready to answer three distinct questions.

First, "What do I want my students to know or be able to do at the end of this teaching segment that they did not know or were not able to do at the beginning?" *What* PERFORMANCE *am I looking for?*

Second, "How will I test them to determine what they have learned?" *What* CRITERIA OR STANDARDS *will I use to measure the speed, accuracy, and/or quality of performance?*

Third, "What equipment or resources will students be given or denied during testing and evaluation?" *Under what* CONDITIONS *will they be tested?*

A properly written performance objective has all three parts: *(1)* the *performance*, or what the student will be able to do; *(2)* the *criteria*, or how well the student will be expected to perform; and *(3)* the *conditions* under which the performance will be evaluated. Although the three parts may appear in any order, it is a good idea to standardize the writing of objectives, to the greatest degree possible, within an individual department or jurisdiction.

Some Examples

1. Given an operational breathing apparatus, wearing full turnout gear, except gloves and helmet,   (CONDITIONS)

   the firefighter will don the breathing apparatus, and be breathing from the air tank,   (PERFORMANCE) performing all steps completely and correctly, in not more than thirty seconds.   (CRITERIA)

2. Given an unlabeled illustration of a standard breathing apparatus, without using notes or reference material,   (CONDITIONS)

   the firefighter will label the parts of the apparatus,
                                                       (PERFORMANCE)

   so that all of the illustrated parts are labeled correctly.
                                                       (CRITERIA)

3. Given a pitot gage and a flowing solid-stream nozzle,
                                                       (CONDITIONS)

   the firefighter will read and record the nozzle pressure,
                                                       (PERFORMANCE)

   to a degree of accuracy within ± 3 psi of the pressure as measured by the instructor.   (CRITERIA)

As we examine the center and the first ring of the training target, writing performance objectives for each segment of information we want to transfer to our students, we are, in effect, creating the skeleton of a course outline. One of the tremendous side benefits of developing and using performance objectives is that we create the basis for testing as we create the outline.

Hints for Writing

In describing performance, use words which specify a clearly understandable, unmistakable action; use words which are subject to the least possible misinterpretation.

Words Open to Few Interpretations	*Words Open to Many Interpretations*
To write	To know
To identify	To appreciate
To sort	To describe
To solve	To understand
To construct	To grasp
To select	To believe

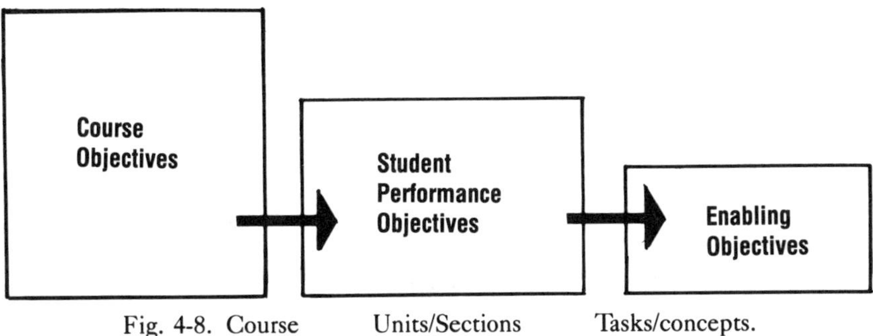

Fig. 4-8. Course Units/Sections Tasks/concepts.

Performance words must describe observable, measurable actions. Mager points out the difference between "overt" and "covert" performance: *Overt* refers to performance which can be directly observed; overt performance is visible or audible. *Covert* refers to mental or invisible performance. Cognitive/knowledge objectives require us to be extremely creative, at times, in changing covert performance to overt performance. Psychomotor (skill) objectives are usually much easier to write.

Levels of Objectives

We keep performance objectives simple and addressed to single concepts by creating objectives at several different levels within courses and lessons. There are at least three levels which you should consider using: Course Objectives, Student Performance Objectives, and Enabling Objectives (Fig. 4-8). Again, the concept is the important thing, not the specific terminology.

Course Objectives are written in very general or "global" terms and usually state only performance, not conditions or criteria. They provide the reader with a broad idea of the subject matter and scope of the course.

EXAMPLE: The firefighter will demonstrate knowledge of the duties, responsibilities, and operations of the engine company.

Student Performance Objectives are written to cover units or sections within the course and usually provide information relative to conditions and criteria, as well as to performance.

EXAMPLE: The firefighter will demonstrate knowledge of the construction, operation, and utilization of fog, solid-stream, and foam nozzles, from memory, without assistance, to a written test accuracy of 70 percent.

Enabling Objectives are highly specific, and they cover the many individual tasks or concepts which must be mastered in order to reach a particular student performance objective. They specify the performance and any conditions under which this particular enabling objective must be accomplished. Enabling objectives may or may not include criteria, depending on how the covering student performance objective is written.

EXAMPLE: Given an unlabeled diagram of a typical variable-flow fog nozzle, the firefighter will identify the major parts and trace the water flow through the nozzle.

In the enabling objective example just given, the standards for evaluation are omitted because the student performance objective covering that unit or section specifies, "to a written test accuracy of 70 percent," and also provides, as testing conditions, "from memory, without assistance." Structuring objectives in this way eliminates a lot of unnecessarily repetitive wording.

Objectives and Motivation

We have mentioned a number of benefits which performance objectives provide. Among the most important is the effect of well-written and clearly understood objectives on motivation. At the end of the last chapter we reviewed three motivating methods and indicated that they were linked to course development: to provide clear reasons why, to provide practical applications, and to provide known objectives. We can end this section by taking those three items in reverse order. Carefully developed and clearly stated performance objectives provide practical applications, and they show both instructors and students how this particular concept or skill fits into this particular course and why it is necessary to the firefighter or fire officer.

The Course Development Process—at Last!

To grow and stay healthy, every fire service agency must welcome new educational ideas and encourage the development of new training programs and instructional materials. Course development and course revision is expensive and time-consuming, however; cost control is essential. Putting these two thoughts together, we must construct a course development process which encourages innovative thinking and at the same time forces us to look realistically at needs and costs. It should be *easy* for any member of the organization to recommend a new training program, but it should be *difficult* for the training division to proceed with any course development project until it has been thoroughly reviewed and the work has been authorized by top management.

Structuring a Course Development System

Let's examine the outline of a typical system[1] and see how to fit these seemingly conflicting ideas together. Keep in mind our earlier suggestion that the outline may contain more steps or fewer steps than you need and should be adjusted accordingly. (Remember "whole-parts-whole"? First let's look at the "whole" system; then we'll go back and talk about the individual "parts.")

1. Submission of instructional idea:
   a. Plan format of submission.
   b. Establish content of submission.
2. Initial review:
   a. Verify as an educational problem.
   b. Research history within agency.
   c. Review operational needs.
   d. Review instructional capability.
   e. Accept (rejection stops process).
3. First administrative review submission:
   a. Develop statement of need.
   b. Write course objective(s).
   c. Establish tentative program format.
   d. Establish tentative program length.
   e. Estimate timetable for development.
4. First administrative action:
   a. Review submission.
   b. Coordinate with external groups.
   c. Approve (disapproval stops process).
5. Front-end analysis:
   a. Prepare occupational analysis.

---

[1] The outline draws from a system developed by Dr. Bruce J. Walz, Thomas C. Cusick, and the author for the Maryland Fire and Rescue Institute.

      b. Prepare target audience analysis.
      c. Analyze instructional requirements.
  6. Second administrative review submission:
      a. Summarize analyses.
      b. Write student performance objectives.
      c. Prepare logistical impact statement.
      d. Establish development timetable.
      e. Estimate development costs.
      f. Establish implementation timetable.
      g. Estimate implementation costs.
  7. Second administrative action:
      a. Review submission.
      b. Coordinate with external groups.
      c. Approve (disapproval stops process).
      d. Authorize funding and resources.
  8. Develop course:
      a. Obtain necessary technical assistance.
      b. Write enabling objectives.
      c. Develop instructor's guide.
      d. Develop student materials.
      e. Develop or purchase audiovisual aids.
      f. Develop evaluation criteria.
  9. Pilot test program:
      a. Successful: implement program.
      b. Unsuccessful: rewrite, retest.
 10. Implement program:
      a. Provide periodic evaluation.
      b. Provide periodic revision.

Four Distinct Stages

The ten steps outlined can be divided into four separate and distinct stages, separated by the three reviewing actions (Fig. 4-9). The three reviews are "built-in" cost-control devices which force us to look at any course development project very carefully, *before* investing any significant amount of time and money (Fig. 4-10). Steps 1 and 2 require only a minimal investment in technical time by the training officer or the training division. Steps 3 and 4 cause us to expend a modest amount of technical effort and a small amount of administrative time. Steps 5, 6, and 7 require a substantially greater number of technical hours and may require some appropriated funding, plus

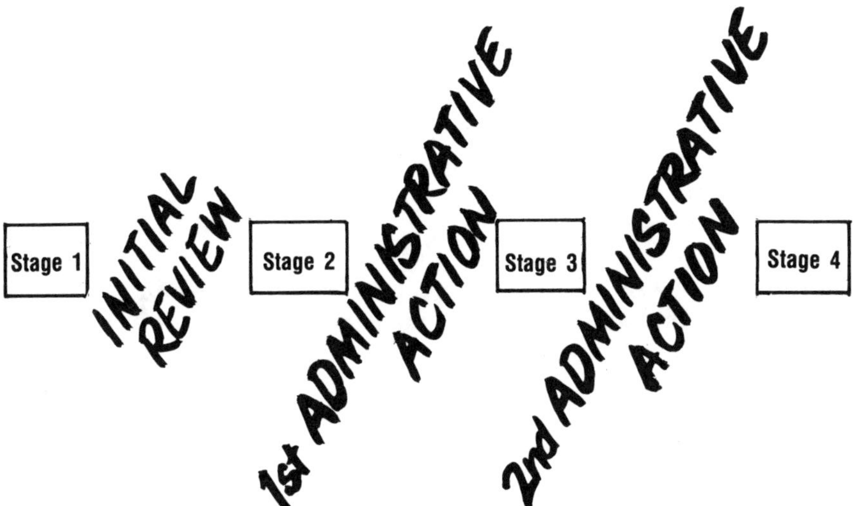

Fig. 4-9. Course development stages.

a close look by the chief administrative officer or his designated representative. Not until steps 8, 9, and 10 are we required to make any extremely heavy investment of time and money; however, any course development project reaching this stage should be of proven merit and have a high potential for success—a much happier set of circumstances than trying to figure out how we happened to be caught up in an ill-conceived course development effort, spending money we don't have, on a subject area of questionable utility—and *that* occurs more often than we like to admit.

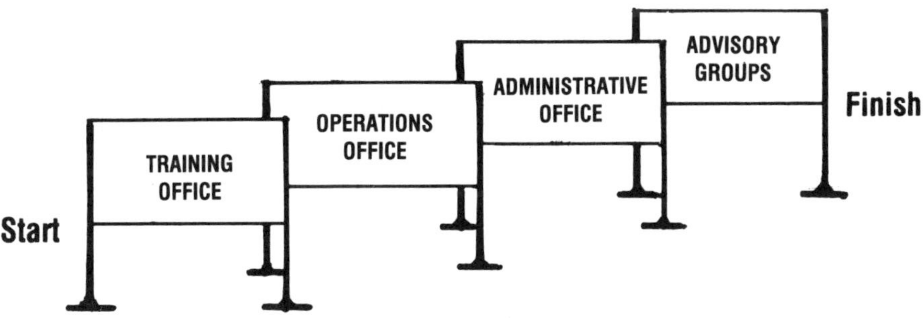

Fig. 4-10. Lots of hurdles.

Filling in the Outline

Submission of an Idea

Step 1 requires us to establish a simple but formal procedure by which any person in the system can submit a course development idea. As we said earlier, it should be easy to recommend a new program but not *too* easy. We recommend a brief, written format; if members feel an idea is worth suggesting, they should be willing to take a few minutes to write it down. A simple one-page form can provide a space for a brief explanation of the subject, course content, or area of interest proposed; space for a statement relating the proposal to a specific emergency service problem or concern; space for a statement on the overall objective of the proposed course; and space for any other pertinent information the originator wishes to submit. We should publicize the procedure and the place for receiving suggestions and provide absolute assurance that all ideas are welcome and will be carefully considered.

Initial Review

Step 2 need not go beyond the training officer or training division. First determine whether or not the submission truly concerns an educational problem. Screen the idea submitted to determine if it is best addressed by education, engineering, or enforcement. If a problem does in fact exist, but you feel it could better be solved by redesigning existing equipment, by purchasing new equipment, or by enforcing existing rules, forward the idea to the appropriate person or division within your system. Send along the original submission, accompanied by a brief written explanation of your thinking, and send a courtesy copy, indicating the action you have taken, to the individual who submitted the idea.

If you agree that the submission is, in fact, a valid educational suggestion, do a little historic research. You should be able to answer the following questions. Have we tried a course like this in the past? If we have tried something similar and it failed, why did it fail? If we tried something similar and it worked, can we bring out the old material and update it to meet current needs? If we have nothing similar in our files, can we obtain such a course from some other fire service source?

If necessary, check with your operational people. Find out from them if they agree that the proposed course has a valid place in the operations of the department or agency; will it help make our operations more effective, more efficient, or safer? Will there by any operational "side effects," which will require us to revise operational procedures or make a heavy investment in equipment? On the basis of a quick review, can they see any particular advantages or disadvantages?

Consider your own instructional capability. Is the proposed program within your capabilities for both development and implementation? Do you have the expertise to prepare and teach the proposed course material or would it be better developed and housed in some other educational agency?

Based on this simple and relatively superficial review, we should be able to say, yes, this is a good idea, worth pursuing further, or no, this does not fit our needs or our capabilities. Rejection at this point stops the process with minimum expenditure of time and effort. In either case, send the person who submitted the recommendation a courteous written response, accepting or rejecting the idea and summarizing the reasons for your action. If you accept, proceed to the next step.

First Administrative Review Submission

Step 3 is the preparation of a submission for review by management. Based on your initial survey, develop a statement of the perceived need. Include as much information as you feel necessary to explain the proposal but, at this stage, keep things relatively brief; you are asking only for approval of the concept and authorization to dig a little deeper into the idea. *Please*, at this point and whenever appropriate thereafter, give credit to the originator.

Write a formal course objective or objectives establishing the educational intent of the proposed course. Establish a tentative course format: classroom study, practical evolutions, or a combination of both.

Almost reluctantly, we suggest that you establish a tentative length for the program. Reluctantly, because too often we set out to write a "twelve-hour course" or a "thirty-hour course" or a course of some other preconceived number of hours, and then we stretch or squeeze the material to fit the time block, rather than adjusting the time block to fit the material. Course length should be determined by how much subject material is in the middle of the training target, *not* by some artificially imposed notion of the number of hours that are appropriate. On the other hand, management has a right to ask how long a program you anticipate, and you need to have an answer.

First Administrative Action

In step 4, you can submit the program to management in writing, make a verbal presentation, or do both. We urge that you do both. Prepare a formal, written proposal, and, if at all possible, review the proposal with management in person. Your chances of obtaining approval are much better in a face-to-face meeting than they are when you rely on a document to speak for you.

Management may or may not wish, at this time, to bring the proposal to the attention of appropriate outside groups. If they do so, you can expect some delay. How long, of course, depends on the circumstances, and how you handle excessive delay depends on *your* relationship with *your* management. We can ask that management review; we can suggest that management approve; but we can't *tell* management to *do* anything. (Perhaps *you* can, but *we* can't.)

At this stage, management disapproval stops the process just short of the point where we need to begin to expend some substantial technical hours and possibly some dollars. Their approval authorizes you to do so.

Front-End Analysis

Step 5 involves what is known as "front-end analysis," "front-end" meaning at the beginning. We have indicated the three most common types of analyses made at this point: occupational, target audience, and instructional.

Occupational analysis involves an examination of the total content of this group of tasks or area of study. In terms of our target, occupational analysis is a study of what the student *must know* and what the student *should know* in order to achieve the objective(s) we have established. In-depth occupational analysis is an extremely time-consuming and therefore costly effort. How deep you go, or need to go, will be determined by the requirements of your own department. We will talk about the technical preparation of a job analysis in the next chapter. Since an occupational analysis is made up of a series of job or task analyses, let's wait until Chapter 5 for details.

Target audience analysis consists of examining, in detail, the student group. You need to know such things as their approximate average educational level, their reading and writing proficiency, their physical abilities, their general level of fire service knowledge, and their approximate experience in the field. Obviously, the scope of this task depends on the size of your department or agency. In a small fire department, "target audience analysis" may consist of little more than sitting at your desk with a cup of coffee, considering what you know about your folks in relation to this particular course. In a large department or in a major county or state training agency, target audience analysis may involve survey forms, computer printouts, and detailed study of a great number of different factors. To begin

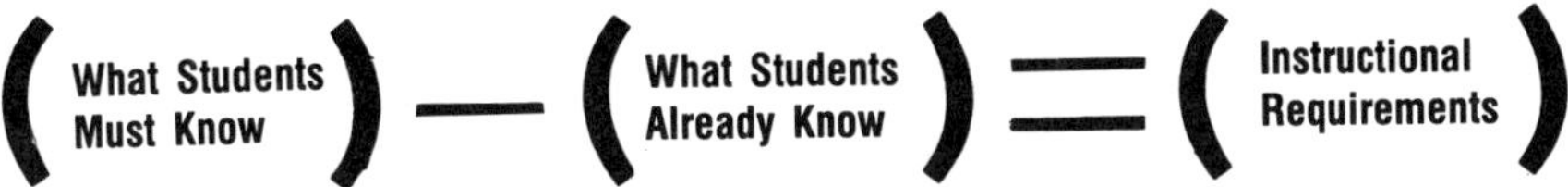

Fig. 4-11. Instructional requirements.

development of a particular course, we need to have a very definite idea of what our people already know in this specific subject area. If you do not have this information, you must establish some means to get it. Valid tools include analysis of training records by course title and hours, conducting educational surveys among potential students, or pre-testing a representative cross-section of your people.

Instructional requirements analysis determines what we need to teach; it assists us in specifying the student performance objectives and the enabling objectives necessary to achieve the course objective(s) we have established. The occupational analysis tells us what students need to know, the target audience analysis tells us, among other things, what the students already know, and the remainder forms the "instructional requirements" (Fig. 4-11). Rather simplified, what the students need to know about this subject, minus what the students already know about this subject, equals the "instructional requirements" necessary for this course.

Second Administrative Review Submission

Step 6 involves the preparation of the second administrative review package. This step requires that we personally be convinced of the merit of the proposed course, because we are going to make some firm predictions on which management will base their final "go-no go" decision. Take your time, work carefully, and double check your figures.

Your chief or the chief's designated representative probably won't be interested in the details of the front-end analysis, but it may be a good idea to prepare a summary and have it available, perhaps as an appendix. If it has been done well, it will provide support for the student performance objectives you write at this stage. In developing these objectives, you will begin to block out the sessions and lessons for the total program. As you work your way through the entire body of subject material, organizing and planning the flow of information and writing the major objectives, you will begin to get a good idea of what the completed course will look like.

Once this framework is complete, consider the resources necessary to finish the job. What is the logistical effect on you or on the training division? Can this work be completed along with other training activities, or must

Fig. 4-12. Do your cost calculations with care.

other tasks slip for this work to be accomplished? Be as honest and as objective as you can. Remember, *you* are the one who has to make the development plan work. If it fails, management will be looking to *you* for the answers. Base your logistical impact statement on facts and on carefully drawn-up estimates, *not* on guesswork and *not* on what you think management would *like* to hear.

Work out the development timetable in as much detail as your organization requires. A milestone chart with starting and completion dates for each major step is an excellent tool. Carefully establishing the major steps will be an asset in estimating the total development costs. You may or may not need to figure "in-house" time, according to the requirements of your agency, but you should include all supplies and services for which purchase orders will be required. Most chief executive officers do *not* like surprises. If you feel it necessary to surprise your boss, we suggest the surprise be in the form of a cost underrun, not the opposite (Fig. 4-12).

Finally, work out your best estimate of the time that implementation can begin. Remember to build in the pilot testing phase and leave yourself some time for revision, as necessary, based on the pilot tests. Pilot tests are not a step you should try to avoid; they are an absolute blessing for training and education staff. Management is almost always looking for production; managers want to see things moving. The training staff is almost always searching for "just a little more time" to polish their creative efforts. Pilot tests provide an opportunity to show management some visible results while you remind them, at the same time, that this is *not* the finished product. Don't overlook the advantage this strategy affords you.

Carefully look at the implementation costs: instructional time, multiple sets of instructional aids where they are needed, student materials, student time (if that is a cost factor), and any special equipment or facilities required. Be as comprehensive and as accurate as possible. This review level gives you an opportunity to display the professionalism of the training staff, whether that consists of one person or twenty. Do your homework well, and construct the review presentation package with style; both will pay off.

Second Administrative Action

Step 7 should unquestionably be a formal, face-to-face presentation to management, backed up by your detailed written proposal. There will very probably be some difficult questions, and you must be ready with well-reasoned and accurate answers. Management again may wish to check with outside constituent groups, and the same rules we mentioned earlier still apply. As before, management disapproval stops the process. Management approval should include both authorization to proceed *and* any special funding required for course production.

Develop Course

Step 8 is where the real work resides. This stage is by far the largest step in terms of effort and the longest step in terms of time. How much effort and how much time, of course, are determined by the scope of the program.

An immediate decision concerns the need for and the selection of outside assistance, if it is required. If you are able to pay for the work, it is possible to "farm out" almost any part of a course development project: the writing of objectives, the writing of the instructor's guide, the writing of student materials, preparation of art work, transparencies, slides, or other instructional aids. Whether or not you do this work "in-house" depends on the time and talents you have available and on the dollars you are authorized to spend.

The logical way to begin is by writing the enabling objectives, followed by the supporting outline or text for the instructor's guide. We show, in outline step 8, enabling objectives, instructor's guide, student materials, and audiovisual aids as four separate development items (Fig. 4-13). In actual practice, they will probably be developed concurrently, with your attention moving back and forth among the four elements, weaving all of them into a

Fig. 4-13. Course components.

coherent whole. Wording which feels right and looks right in the instructor's guide will suddenly seem awkward or unnecessarily complex when it appears in the student materials, forcing you to change one or the other, or both. Each word you, or your colleagues, write in one area will ricochet among all the other areas, causing changes in wording or changes in approach. It is a totally dynamic process, and it will remain dynamic until the final editing of the final product.

If you like detail work, "dotting i's and crossing t's," you will enjoy course development. If you do not like detail work, course development will just flat drive you crazy. It takes rare skill and rare patience to produce a course package in which all the parts—instructor's guide, student materials, and visual aids—match in every detail. Making everything consistent plus producing a correct and accurate system of outline headings, eliminating all misspellings and all incorrect grammar, plus ensuring that all technical content is accurate and up to date is a job that will make you grow old before your time. It's fun, but it ain't easy.

You may be thinking—where are the lesson plans? At this stage, they are hidden in the instructor's guide. We feel, quite strongly, that some serious confusion has developed in the past few years between instructor's guide lesson plans and the kind of lesson plans which *should* be used in teaching at the front of the classroom. We will discuss lesson plans and the instructor's guide and how they fit together in Chapter 5. (Trust us. We won't forget.)

The final segment of step 8 is to produce the "evaluation criteria": any necessary quizzes and the final test. As we mentioned earlier, performance objectives provide us with major assistance in this area and, if we have written our objectives properly, test development should be a relatively simple process when the rest of the course is complete. We must, though, put the written questions in correct form, properly paged, keyed to the reference material, and we must develop an accurate and equitable performance testing procedure where one is required. We will go into detail in Chapter 9.

Pilot Test Program

Step 9 provides us with the opportunity to see how well we have done. A pilot test is conducted by teaching the new course in as close to finished form as you can make it. There is a lot of flexibility in just what "finished form" means, but be careful in the amount of liberty you take. It almost always makes more sense to use copier-produced materials than to go to print before the pilot tests are complete. We sometimes use quickly produced interim visuals, transparencies in particular, which may be of lower quality than the final product. Experience indicates, though, that we pay a penalty for going to pilot testing with a product that is "quick and dirty" rather than one that is reasonably polished and finished. This penalty we pay is inaccurate assessment of the course. The success or failure of a course of instruction is based on the sum total of the success or failure of *all* of the course components. If too many of those components are not in final form, you cannot obtain an accurate assessment of the final product.

If time permits, one solution may be a *series* of pilot tests in which materials are improved in stages as the final design gradually becomes frozen. Use your common sense; there is, once again, no magic formula. Course development is an art, not a science.

Another pilot test decision area is in the selection of the student groups. You may want to consider conducting the initial pilot with a group of instructors sitting as students. This audience can help you assess both the instructional organization and the technical content of the course. Putting instructional staff through a new program is also an excellent way to prepare them to teach the material. Actual student trainees can be selected by name or chosen at random. If you select the group, use care not to assemble either all superior or all inferior learners; pick a representative cross-section. A random selection process may give you a cross-section of learners but random selection can also produce a group skewed high or low.

There are so many variables that we could spend more time on the topic than it is worth. Just be careful. Think objectively about the pilot process you are using, and think objectively about student response. When we, as instructors, have spent months of hard, detailed work on development of a course package, it is difficult to be truly objective concerning criticism of our efforts. Establish the pilot test program as carefully as you can, conduct the course as well as you can, using the best materials you can put together, and assess the student response as objectively as you can. If you can accomplish those aims, you will be doing *your* job better than most.

Based on pilot test results, we have a number of options. We can implement the program, we can revise and then implement the program, or we can revise and retest. No matter how well we have done our work, at least some changes will usually be necessary as a result of actually teaching the course. If these changes are major, retesting may be indicated. If the changes are minor, and the overall evaluation of the course is favorable, it may be perfectly acceptable to package the course in final form and push it out the door.

Implement Program

Step 10 is ongoing and long-range. With a totally new course, it may be well to evaluate test results and instructor and student opinions in six months or a year. When a course becomes established, a longer cycle is appropriate. Just how frequently periodic evaluation and periodic revision should occur depends on how many programs you must monitor and the personnel and resources you have available to do the work. Most fire training agencies seem to aim at about four or five years for major course revisions. The key consideration is to develop a system which will ensure that periodic evaluation and revision *will* be done.

One effective method is to put each of your programs into a revision calendar cycle of four or five years. As suggestions or comments on any particular course are received, they should be noted and filed in a book or folder for reference at the next revision cycle. When a course becomes due for revision, the file is pulled and serves as the basis for the revision effort. If a really serious problem comes to your attention, such as discovery of a significant error in the material or perhaps a major change in the technology, you can issue a handout or add addenda or errata sheets to the course material. Again, the formality of the process depends on the formality, and the size, of your organization.

The depth of your course revision process will depend on the scope of changes a particular program requires, and on the age and the continuing utility of the course material. It may be a good idea to use the course development outline as a general guide, not necessarily following it in detail, but checking to see what steps might make sense in a revision. One item you might want to consider is checking with management to see if the course remains consistent with any long-range planning currently under way or to see if management feels the need to get in touch with any constituent group as part of the revision process.

Fig. 4-14. Dear Aunt Jane, . . .

Course Development and Giraffes

We once read about a little girl who sent her aunt this thank-you note for a birthday gift: "Dear Aunt Jane, Thank you for the giraffe book. It told me more about giraffes than I really wanted to know." We have a vague feeling that we may have told *you* more about course development than you really wanted to know (Fig. 4-14).

We have spent a considerable amount of time here because the fire service *talks* a lot about course development but very seldom *does* any. Done well, the process is an extremely difficult and exacting job. If you are already involved, perhaps what we have written will be of assistance. If you are just beginning, we hope we have given you enough ideas to get things up and running. Good luck!

Let's move on to lesson planning, and we'll tell you why we are convinced that instructor's guide lesson plans and teaching lesson plans are *not* the same thing.

5

Lesson Planning

Back when sodium bicarbonate was the only choice in dry chemical, a "lesson plan" was created by the fire instructor for strictly personal use. In those days, our lesson plans were "home-made" simply because there were no others available. We produced highly individual documents, full of personal shorthand, symbols, color codes, arrows, and scribbles, with sections crossed out and notes stuck in the margins as we learned what would work and what would not. Most of these would have been absolutely useless to another instructor, but they were excellent for the person who wrote and used them. We believe this is *still* the type of "lesson plan" you should take into the classroom.

The problem we referred to in Chapter 4 stems from the fact that today there are all sorts of teaching guides available: some developed within fire departments, some produced by county, state, or federal fire training agencies, and some marketed commercially. While we are delighted to see so much excellent fire service training material being created, we feel that ready availability of these "canned" programs has led many new fire instructors to attempt to teach directly from these typed or printed instructor's guides. That practice, we believe, is totally wrong.

Once more we bump into terminology. We are going to use "instructor's guide" to designate lesson plans produced by any central office, either public or commercial. An instructor's guide, therefore, is a centrally produced, standardized body of educational information, outlining the established ob-

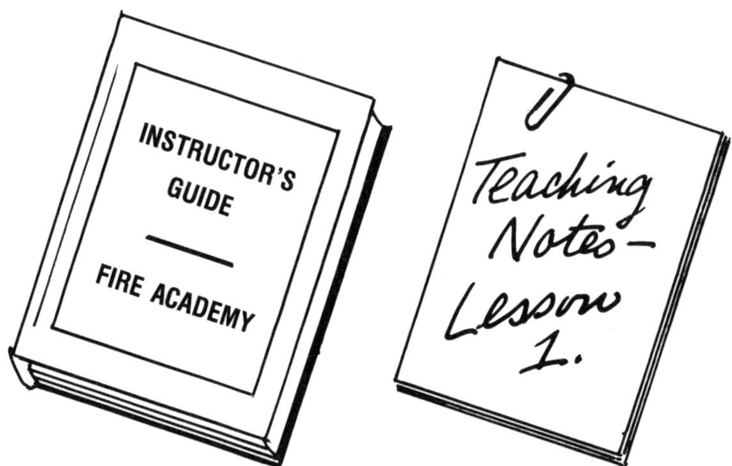

Fig. 5-1. Different in purpose, format, and appearance.

jectives, the desired content, and the teaching sequence recommended or required. We feel very strongly that instructor's guides are designed to tell the instructor *what* to teach. They are *not* designed to teach from, in the classroom.

We are going to use "teaching notes" to designate the lesson plan you should have before you on the lectern in the classroom. Your teaching notes are your personal plan for covering the points necessary to achieve the performance objectives established by the instructor's guide (Fig. 5-1).

There are then, in our opinion, "instructor's guide" lesson plans and "teaching" lesson plans, each containing and covering essentially the same information but significantly different in purpose, format, and appearance.

Instructor's Guides

Let's begin by talking about formal instructor's guide lesson plans. These, as we said, are developed centrally and establish a standardized, reproducible plan for teaching the stated objectives of a particular course. Within the course guide there will be one or more lesson plans for each teaching session, for each block of time the instructor meets with the students. Each of these individual lessons will contain a number of specific sections, arranged and organized in a standard format.

If you are charged with the responsibility for developing instructor's guide lesson plans in your department or agency, you should establish a format suitable for your organization and, to the greatest degree possible, keep to that arrangement in all of your courses. Establishing a reasonable level of standardization will greatly expedite your course development work, and it will also be helpful to the users of your lesson plans.

There is no one "right" arrangement for a formal instructor's guide; there are probably as many variations as there are agencies producing lesson plans. Most, however, include an administrative heading, an introduction, a major block containing the teaching presentation and application steps, testing information, and a summary or closing. We will outline a typical format for organizing sequence and content. Once again we must urge you to consider our suggestions as general ideas to be tailored to your needs.

Administrative Heading

Each instructor's guide lesson plan should begin with an opening section or heading which provides necessary administrative information for the user (Fig. 5-2). This page, or two, if necessary, should contain:

- The course title
- The session and/or lesson number
- The lesson topic
- The level of instruction intended
- References
- Training aids and equipment required
- Estimated teaching time required

Course Title

The formal name of the overall course of instruction of which this session or lesson is a part is called the course title.

Session/Lesson Number

The session/lesson number is any numeric or alphanumeric designation which meets your needs. Since student attendance is ultimately scheduled according to clock hours, your lesson planning, and therefore your lesson cataloging system, probably should be built around some standard block of time.

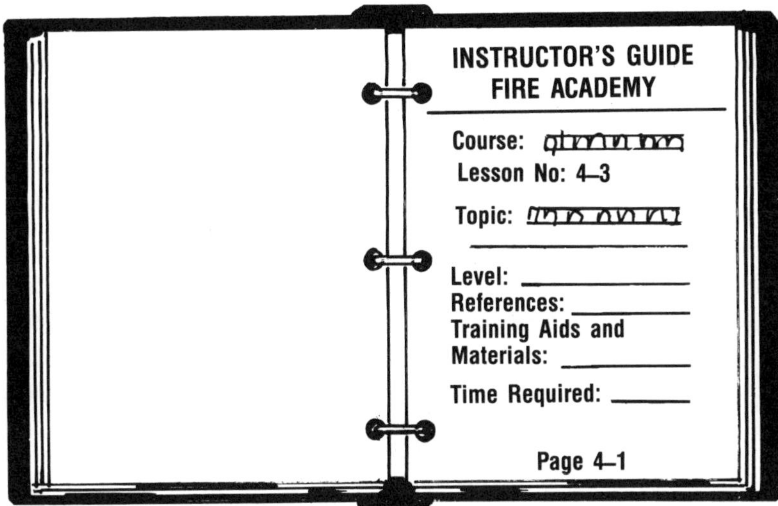

Fig. 5-2. Administrative heading.

One simple yet effective system, developed by Thomas C. Cusick for the Maryland Fire and Rescue Institute, designates a "session" as a three-hour block of teaching. Within each "session" there are three one-hour "lessons." The session, lesson, and objective numbers form an integrated alphanumeric sequence. Session 4, for example, is the fourth three-hour block within the course. Within session 4 are three one-hour lessons; 4-1, 4-2, and 4-3. Each lesson covers one student performance objective (SPO) which carries the same number as the lesson: lesson 4-3 covers SPO 4-3. The enabling objectives (EO) which support that SPO follow in the lesson as EO 4-3-1, EO 4-3-2, EO 4-3-3, and so on. The system is useful, it is logical, and it is very easy to follow. If necessary, of course, these basic divisions can be manipulated to accommodate the odd-sized block of material. There is no reason why a lesson cannot be expanded to two hours, or cut to half an hour. The time frame for the lesson should accommodate the complexity of the objective.

Your own system may be as simple as "lesson 3," or as complex as one of the military designations, such as "NAVPERS 16103-C, 1.4.2A." If you are just beginning to catalogue the course offerings of your agency, keep the system as simple as possible, but be sure to use a logical sequence and allow plenty of room for future expansion. Cataloguing systems never grow smaller; they only grow bigger.

Lesson Topic

The lesson topic is a brief descriptive title or phrase indicating the subject area or task covered in this particular lesson.

Level of Instruction

The highest level of information the lesson requires or is intended to reach is called the level of instruction. This numerical designation provides the instructor with some idea of the level of performance which will be demanded of the students and establishes an idea of the type of instruction which will be required. It serves as a helpful "bench mark" for the using instructors.

References

References constitute the texts, manuals, standards, or codes on which the development of this lesson was based. Be sure to include edition numbers and/or years of publication. These data provide the instructor with original source references and also serve as a reminder that periodic updating may be needed as the original source materials change. This section should *not* be a bibliography of recommended material which students are expected to read; that information should be located in the body of the lesson or in a separate student handout.

Aids and Equipment

This element—aids and equipment—should be a functional list of everything necessary to teach the lesson. It can be comprehensive, listing every individual item required, or it can be abbreviated according to local custom. Just be sure it is *meaningful* to those who need the information. Any instructor within your system who reads the aids and equipment list should know exactly and completely what must be ready and available to teach this particular block of instruction. Include such items as films, slide sets, transparencies, projectors, training aids, handouts, and the tools and equipment necessary for both the instructor and the students, all described at a level of detail which *your* instructors will comprehend.

Estimated Teaching Time

The last item in the administrative heading is your best estimate, in

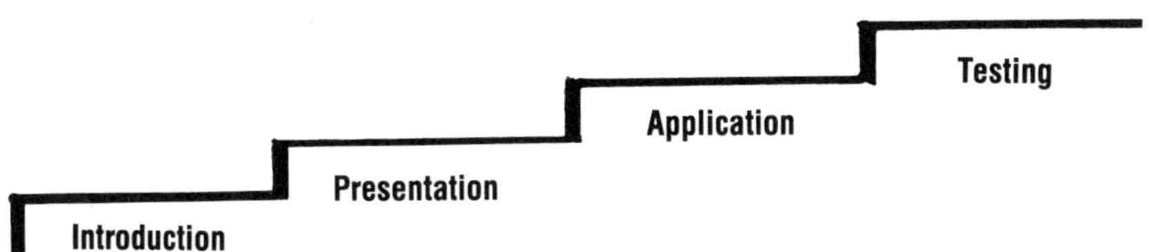

Fig. 5-3. Four-step method.

minutes or hours or both, of the time required or available to teach this session or lesson. Remember to consider break time; a three-hour session minus two ten-minute breaks equals only 160 minutes of teaching time.

The Body of the Lesson

In any lesson of instruction, the content to be transmitted to the students, along with the evaluation process, is divided into four main sections:

- Introduction
- Presentation
- Application
- Testing

Together, these four sections are often designated as "the four steps" or "the four-step method" of teaching (Fig. 5-3). The concept has been in use at least since World War II, but it is still perfectly valid and useful for those of us involved in day-to-day fire service instruction.

Introduction

This first step involves only a small percentage of the total lesson time, but it is absolutely essential and extremely important to the success of the rest of the teaching process. Properly executed, it spells out for the students: "This is what we are going to cover, and this is why you should be vitally interested."

The introduction section, sometimes called preparation, should provide for the instructor:

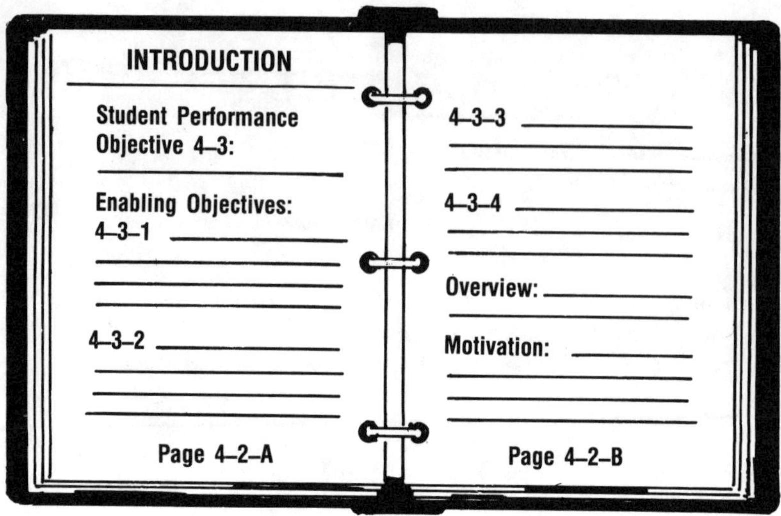

Fig. 5-4. Introduction.

- The student performance objective for this lesson
- The enabling objectives required
- A brief narrative overview of the subject area (the "whole")
- A block of recommended motivational suggestions designed
to assist the instructor in capturing the interest of the student group

In the same process, with the same words, you as the course developer are trying to awaken the interest of the instructor, to convince the instructor that this lesson covers a truly important subject area, worthy of the best possible presentation (Fig. 5-4). It is a place to be creative. Apply the motivational points we talked about in Chapter 3 to establish a foundation which is consciously positive, up-beat, and enthusiastic.

The introduction is *not* an option. It is an integral and essential part of every lesson. In the lesson planning process and in teaching this section in the classroom, give it the time, the attention, and the emphasis necessary to capture the imagination and the interest of your students.

Presentation

The name of this section was well chosen. Here the instructor "presents" to the students the subject material that the lesson is designed to transmit. *Presentation is the heart of the lesson.* It is the section in which we introduce and display and describe the facts, the ideas, the concepts, the processes, the skills we wish to convey. It is where we *teach.*

In the development of instructor's guide lesson plans, this section is the area which will require the most effort and the most time. Ensuring that the content is accurate, logical, and absolutely clear requires a lot of work and a critical eye. It is not a process that can be or should be hurried.

Decisions, Decisions, Decisions . . .

In the construction of the lesson presentation, there are all sorts of decisions which must be made. Some of them you will be forced into as the work progresses; others you can anticipate. The greater the effort you put into planning and looking ahead at the entire project, the easier your work will be.

Lesson Scope and Content

If the course development process was effectively done, you should have a definite idea of what must be covered. What the students need to know about this subject, minus what they already know, equals the instructional requirements or course content. Course content divided by sessions equals session content. The length of the program and the number of sessions or lessons may expand or contract as we get into the detailed work, but we must have, at the beginning, a good idea of content limits. Without this basic information, there is a good chance you will leave out some critical section and a better chance that you will wander on far beyond the intended scope of the course.

Arrangement of the Lesson Points

Back in Chapter 3, we talked about the importance of logical sequence and arrangement; this is the point at which sequencing and the flow of information become top priorities. Earlier in this chapter we suggested that the major subdivisions of the lesson plan outline should be the enabling objectives for the lesson. Beneath each enabling objective you can use any classic or standard outline format.

For example:

    I. Major teaching points
      A. Major subsections

> 1. Minor subsections
>   a. Specific points of information

In general, keep the outline as simple as you can. If you go much beyond the four divisions suggested above, you will find that multi-level, sub-sub-sub points serve only to complicate your development task and are extremely difficult for the using instructors to follow.

Assuring a Logical Flow

Within the structure of the outline, select for each section a logical framework on which you can arrange the necessary information. Some of the options are:

- *Chronological:* for example, the sequence of events in a standard operating procedure
- *The required order of actions:* for example, the order of steps in setting a sprinkler valve
- *The order of importance:* for example, the points in a decision-making process
- *The order in which students will encounter or use the information:* for example, a series of components in a hydraulics equation

Which *basis* you choose is not particularly critical, so long as the information flows comfortably. What *is* critical is that you *have* a logical basis which will be readily apparent to a knowledgeable reader. The using instructors must be able to understand and to follow the flow of information, and they must be able to explain the rationale behind the sequencing to their students, if it is questioned. If you draw a blank in trying to establish some sensible arrangement for the material you wish to present, remember "whole-parts-whole," or consider arranging the lesson information from simple to complex or from known to unknown. Don't get "locked in" too quickly. Take your time, do your early work in draft form, and be willing to modify your original ideas if better ideas develop as the work progresses.

Methods of Instruction

You will also, at this stage, find it necessary to decide on the teaching methods you wish to build into the instructor's guide. We will talk at length about methods of instruction in the next chapter; at this point, we will only mention that you cannot develop the presentation section of the instructor's

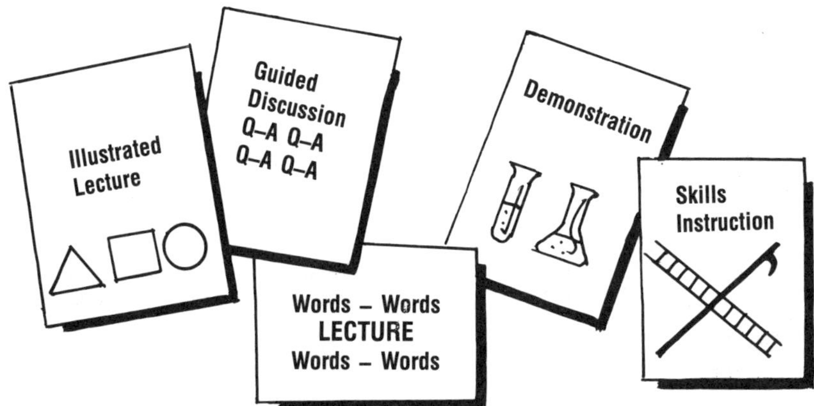

Fig. 5-5. Methods of instruction.

guide lesson plan without making conscious and firm decisions as to how this particular portion of the material can best be presented in the classroom (Fig. 5-5). You will need to consider such questions as the type of information to be taught, the qualifications and the talents of the using instructors, the knowledge and experience of the students, the number of students anticipated per class or section, the audiovisual equipment available, and your budget allocation for developing or purchasing teaching aids and student materials.

Instructor's Guide Page Arrangement

If at all possible, you should work *from the beginning* in the page format and layout you will use for the final reproduction of the instructor's guide. This practice, obviously, eliminates the need to restructure or reprocess the words you write. We will talk about computers in more detail later, but we must say, at this point, that word-processing systems and computer equipment are not luxuries to course and lesson plan developers; they are essential tools of our trade (Fig. 5-6). Any one of the more versatile personal computers and any one of the good word-processing systems will do the job. If you are making a career in training, insist that your agency provide the necessary hard- and software and teach yourself word processing. It will absolutely pay for itself.

As in all aspects of lesson planning, there is no one single page arrangement which is preferred above all others. One page layout, which seems to work very well, was developed by the Maryland Fire and Rescue Institute

Fig. 5-6. Tools of the trade.

and has been adopted by all of the major fire training agencies in that state. The presentation and application portions of MFRI instructor's guide lesson plans are arranged on facing 8½-inch by 11-inch pages of a standard three-ring binder. Each page contains two vertical columns. Reading across the top of the two facing pages, the columns are headed:

Enabling	*Main*	*Instructor*	*References*
Objectives	*Facts*	*Notes*	*and Aids*

This arrangement displays for the using instructor all the necessary information for each point in the outline as it unfolds (Fig. 5-7).

The column headed "Enabling Objectives" contains just that, the enabling objectives, in proper order, for this lesson. For clarity and convenience, you may wish to note in this column any major teaching points, within the enabling objectives. Under "Main Facts" is the lesson outline, the body of material to be taught. The "Instructor Notes" column is used for any points you wish to bring to the attention of the instructor, items which make the teaching job easier or more meaningful. Specific applications of the principles being taught, teaching techniques, suggested class activities, definitions, explanations, or expansions of the outline material are all examples of points which may be inserted in this column. "References and Aids" is the place to list the specific publication and page references you wish to include; it also provides space to identify the teaching aids which are to be used at that point in the lesson, the names of films or videotapes, the code numbers of slides or transparencies, or any similar pertinent information.

In preparing lessons in this format, a critical aspect requires carefully

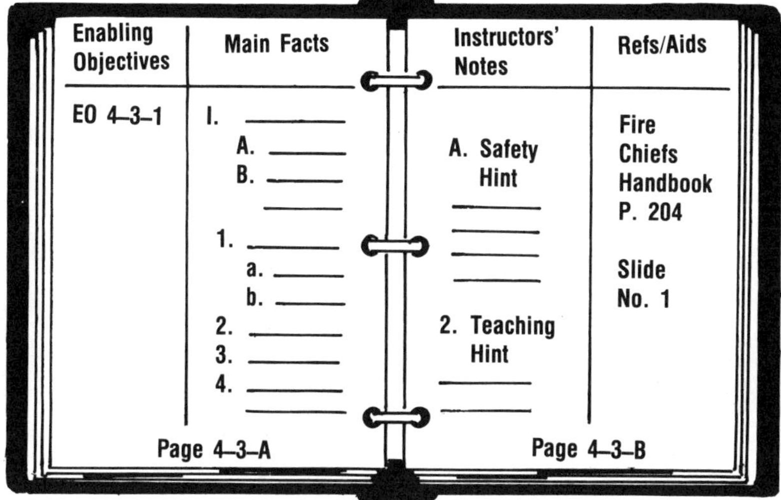

Fig. 5-7. "Presentation/application" page format.

registering the vertical alignment of items from column to column and, in particular, from page to page so there will be no question as to which items belong together. It is a time-consuming and frustrating task, but it is absolutely necessary. The developer must *ensure* that there is no point in the program where the using instructor will lean back and say, "Now, how the heck does *that* fit in?" Again, the process is not a task that can be rushed. Patient, painstaking work in the development and layout stages will, without question, save everyone concerned a great deal of time and frustration later on.

While we are on the subject of clarity, it is generally a good idea to avoid crowding, in laying out an instructor's guide, to leave a lot of "white space" on each page. An open format will look cleaner, and it will be much easier for your instructors to follow.

Skills Instruction Lesson Plans

Skills instruction lesson plans are generally easier to construct than lesson plans concerning knowledge or cognitive subjects. Skills, in and of themselves, flow logically from step to step and from point to point. Our major difficulty, as developers, is to determine how much detail to put into the lesson plan. How much do we specifically include and how much do we leave to the discretion of the using instructor? How do we ensure that the

Job Analysis Form

Course: _______________

Skill Area: _______________

Job: _______________

Main Steps	Key Points
1. _______	1. _______
2. _______	2. _______
3. _______	3. _______
4. _______	4. _______
5. _______	5. _______
6. _______	6. _______
7. _______	7. _______

Fig. 5-8. Job analysis form.

instructor will teach the skill exactly as we want it taught? We must include every step and every point which is *critical,* every item students need to know to perform the skill. Job analysis is the tool we use to make sure all the essential main steps and key points are adequately covered.

Job Analysis

Job (or task) analysis is a relatively simple but very time-consuming process. It requires patient, careful work and a willingness to examine critically and agonize over details—not everybody's cup of tea.

A typical job analysis form has, at the top, space to identify the job under study and the relationship of this job to the entire subject area or field. As always, make sure the job analysis heading contains the information necessary to meet *your* needs (Fig. 5-8). The body of the form is divided into two equal vertical columns. We prefer column headings of "Main Steps" and "Key Points" as simple but clearly descriptive titles. As we work our way through the job, "Main Steps" are the things which the learner *does.* "Key Points" are the things which the learner should *know.*

Job analysis is simply a matter of slowly, carefully, and comprehensively examining and recording the entire sequence of actions involved. As each action step is recorded, any necessary key points are noted. Key points include safety measures, reasons "why," exceptions, "tricks of the trade"—

any factual information you feel is important to the successful accomplishment of the task. Job analysis is definitely not a solo operation. It should *not* be performed at a desk merely by thinking about the task. Get a sharp crew together, people who are interested and who can intelligently contribute, and actually do the work. This procedure will ensure that you cover the entire process, and it will probably develop ideas which you might miss if you were working on your own. Two or three or four heads, carefully selected, will be far superior to one.

How Fine Do We Grind?

The most difficult aspect of job analysis is deciding how deep to go, how specific to be in dividing and defining the action steps. The two extremes, of course, are being too general or being too specific. If the steps are too broad, they may be ambiguous and either difficult or impossible to follow. If you are too exacting, detailing every position, every tiny motion, it wastes development time and is cumbersome to follow. We certainly don't have any magic formula to recommend; as the old saying goes: "Circumstances alter cases." You will probably find it necessary to be more specific in some areas and less so in others, developing a feeling for what is right through your own experience.

We have heard it said that you should do an in-depth job analysis for every task you teach. We seriously question that advice. Job analysis, properly done, is so time-consuming that we simply don't feel it is possible for most fire service training organizations to be that comprehensive. There are, however, at least two occasions when it will be particularly helpful: *(1)* when you must develop a lesson plan for teaching a new technique or the use of new equipment, and *(2)* when you or your instructors are encountering problems in teaching a particular skill. In those circumstances, job analysis is essential.

Some Final Words on Developing the Presentation

In developing a formal instructor's guide, try to keep in mind the viewpoint of the *user.* Examine each section as it is completed and go back and review the entire document from time to time as the work progresses. See that your words say what you mean to say. Make sure that each teaching point will be identifiable and understandable to the using instructors (Fig. 5-9). Be certain that transitions from point to point and from subject to subject are clear and definite. It makes sense to have the material read and critically reviewed by several competent colleagues. A second and third pair

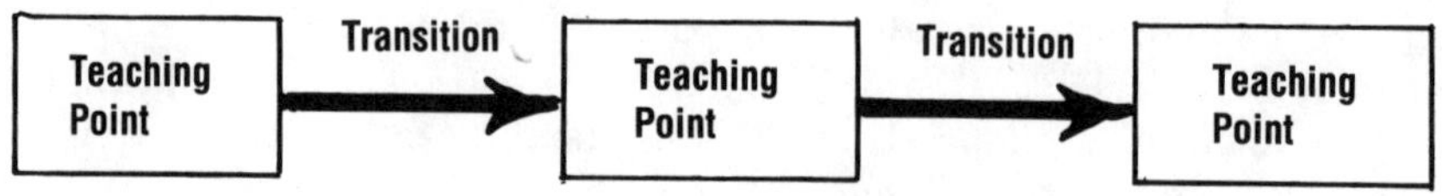

Fig. 5-9. Absolute clarity is essential.

of eyes will see things from a fresh perspective when the trees begin to get between you and the forest.

Application

Within the lesson guide, the application step indicates for the using instructors the activities you, as the course developer, are mandating for the students as a part of the learning process. It is the part of the lesson where students personally apply the information they are learning. Because of the importance of individual proficiency in our profession, application may, in some areas of fire service instruction, absorb the largest percentage of class time.

Application provides an opportunity for students to become actively involved in the learning process, to use the principles and concepts learned in cognitive subject areas, or to practice physical skills. The transition from "presentation" to "application" may be definite and very clear, or it may go almost unnoticed as the instructional process shifts from demonstration (part of the presentation step) to active student participation (application).

In skills instruction areas, application is provided by having students perform the tasks under supervision. We will go into more detail under methods of instruction, but the process, simply stated, is: student performs; instructor checks for correct motions, correct sequence, and correct safety practices.

In areas involving knowledge and mental skills, application can be provided through a wide range of problem-solving exercises. Paper problems, calculations, case study analyses, tactical simulations, building surveys, and construction of pre-fire plans are all part of an almost infinite variety of ways to put cognitive learning to use. We will take a closer look at some of these in Chapter 7.

Testing

While evaluation is considered a part of the four-step teaching process, we will also separate it and cover it in detail in a separate chapter (Chapter 9).

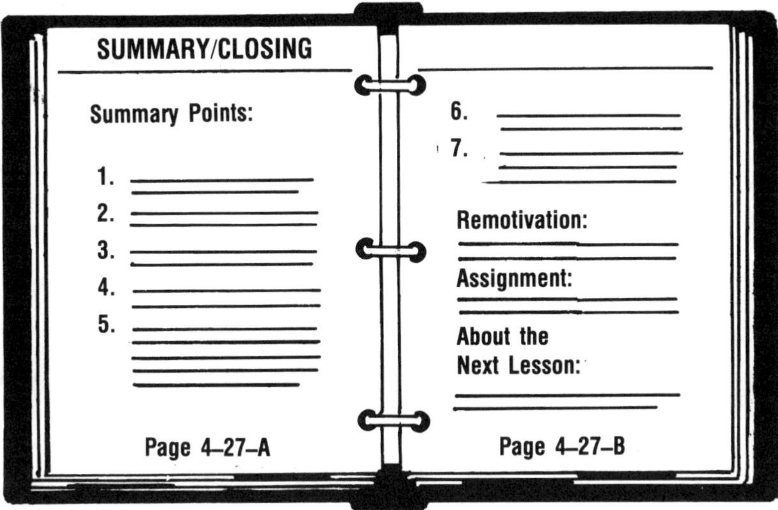

Fig. 5-10. Summary/closing.

It is usually necessary to integrate a variety of formal testing and evaluation tools into the teaching package in the form of graded problems and projects, quizzes, and mid-course and final examinations. Remember performance objective criteria? This stage is where all our hard work in developing well-constructed performance objectives will pay off.

We will leave the subject at that, for now. Please remember, though, that the completed instructor's guide must indicate the points at which you want formal evaluations to occur, and the course development package must include all the necessary quizzes, tests, and any other evaluation instruments you choose to employ.

Summary Closing Section

As the formal lesson plan began with an administrative section, it closes with a segment which sums up the teaching activity and prepares the students and the instructor for the following meeting (Fig. 5-10). The instructor's guide should include a summary outline of important points you want the instructor to cover. This section is seldom very long—usually a page or less. It serves primarily as a reminder to the instructor of the items to review with the students in the closing minutes of the class.

This section also includes any assignments the students are to carry out before the next class meeting: pages of text to be read, problems to be solved, projects to be completed, or any similar activities you want them to accomplish outside of class time.

Other summary/closing options you might want to consider are a re-motivation section, in which the importance of the lesson just completed is re-emphasized, or a few sentences about the upcoming lesson to point the instructor and the students in the direction they will next be traveling.

Teaching Notes

Back at the beginning of the chapter, we spoke of the difference we perceive between the formal instructor's guide and the teaching notes which we feel should be used in the classroom. There are a number of reasons why we urge you *not* to teach directly from the instructor's guide.

Print Size

First of all, in typewritten or printed form, the words in the instructor's guide are usually too small to be seen unless you stand right at the lectern. This fact encourages reading directly from the guide, and it ties you too tightly to one spot and to the guide itself. We feel it is essential for the instructor to possess the freedom to move about the classroom, among the students, to establish the rapport which is so essential in good teaching.

Formality

Second, the language of such a guide tends to be overly formal; instructor's guides are seldom written in the casual, relatively informal manner in which we speak. If you try to teach directly from the guide, you will tend to use the language you see on the written page, and the words you find yourself saying will not be *your* words but the words of someone else. Nobody speaks and writes in exactly the same way, so even if you *wrote* the instructor's guide, and the words are your own, the words will still not necessarily flow well in spoken form.

The Need for Personal Preparation

Third, human nature being what it is, if you work directly from the instructor's guide provided by the agency, you will do less to prepare yourself to teach than you will if your create your own teaching notes, your own personal lesson plans, from the formal guide. In our experience, if you pick up a formal lesson guide from which an instructor is teaching in the classroom, the first lesson or section will be fairly heavy with personal notes, the second lesson will have fewer, the third fewer still and somewhere near the middle of the document they will trail off to nothing. It is just too easy to rely on what you have been handed, too easy to slack off on personal planning and depend on the "canned" lesson material and visual aids to do the job for you. In our estimation, that is not teaching.

The Need to Adapt

Finally, and probably most important, while the instructor's guide establishes the objectives and describes lesson content, it cannot account for the subtle but very real differences which exist from class to class, from group to group. You, in the role of instructor, face to face with a particular student group, will often find it helpful to modify the order, the emphasis, or perhaps even the *content* of the lesson.

Heresy? Anarchy? Not at all. Every competent instructor consciously or subconsciously modifies formal lessons to accommodate the strengths, the weaknesses, the interests, and the needs of individual classes. The best instructors teach to the desired objectives, and they cover all the necessary points, but they may not, indeed they probably *will* not, dot every "i" and cross every "t" in the instructor's guide. And to that we say, "Hooray!" That's what instructors are for. We are not, and should not be, teaching machines. Slavishly following the instructor's guide will not make you a better instructor; it may do exactly the opposite.

Back to Basics: Create Your Own Lesson Plans

When you, as a teacher, are issued a formal instructor's guide, read it through, study it, study the student handouts, the visual aids, and all the other components of the teaching package. Really dig in and analyze what you have been handed. What are the strong points? Where is it weak? Based on your experience with your students, what parts will work well and what parts appear to be problem areas? How does it flow? Does it feel comfortable, considering your experience and your knowledge of the people you will be teaching? When you feel you know the material almost as well as the person who produced it, we urge you to write out your own individual *teaching* notes, *your* lesson plan, complete with the personal shorthand, symbols, color codes, arrows, and scribbles we mentioned earlier.

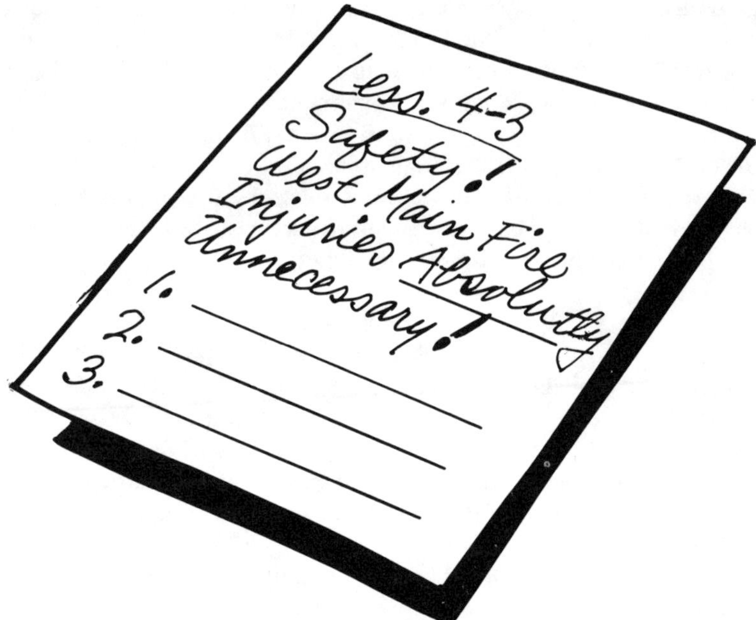

Fig. 5-11. Personal teaching notes.

Personalized Format

The appearance and format are entirely up to you. We prefer single, loose sheets of 8½-inch by 11-inch notepaper, *not* stapled together and *not* in a three-ring binder. Stapling and binders force you to flip pages, and they both hide what is coming next. By keeping your teaching notes in loose sheets, you can display two, side by side, on the lectern, reading, of course, from left to right (Fig. 5-11). As you come to the end of the left-hand sheet of notes, slide the right-hand sheet over the left, uncovering the next page. In this fashion you *always* have the upcoming page in front of you, and you can glance ahead and see what twists and turns you are approaching in the lesson. This technique is extremely helpful in keeping the flow of the lesson moving smoothly, in integrating aids with minimum disturbance, and in thinking about how you will make subject area transitions or work in a break when one is needed.

We keep the loose sheets for a particular lesson clipped together in a manila folder which has, on the inside of the front cover, a list of student handouts, visual aids, and any other items or materials necessary to teach that lesson.

Personalized Printing

Each of us writes or prints in our own distinctly individual style. Use whatever works for you in the classroom and whatever is easy to lay out on paper. We print in large block letters, using a soft pencil—pencil for quick correction of errors, and soft so the lettering is heavy, black, and easy to see. The large block letters are visible from six or eight feet away, which enables us to follow the lesson sequence without standing right at the lectern; a glance in that direction is enough to keep us on course. Don't crowd your words; paper is cheap and space separates and sets off your various items and notes.

Personalized Shorthand

One of the tricks in making the transition from a formal instructor's guide to personal teaching notes is to reduce lengthy descriptions or concepts to just a few words. Absolutely avoid writing in detail all the words you intend to speak. None of us can teach out of a book. If you try, you will end up reading, and that is the *worst* possible classroom situation. To teach effectively, you must *know* the material so thoroughly that just one or two words will trigger in your mind the entire concept or sequence you will next deliver to the class. And you must be able to deliver it on your feet, eye to eye with the class, *not* with eyes locked to the document in front of you.

This does not mean you cannot read a quotation, nor does it mean you should not write details into your teaching notes where they are necessary. It does mean you have a lot of work to do *before* you enter the classroom.

A personal color code can be used to underscore or circle or print, to highlight specific lesson plan elements. Use one color to flag visual aids, another color to indicate handouts, perhaps a third to highlight prepared questions you want to ask. The number of colors and how you use them is entirely up to you.

It is not the least bit important that you follow our specific suggestions concerning format and style. Your own ideas may well be better than ours. Use your imagination and your creativity to develop personal lesson plans or teaching notes that best support you and your style of teaching in the classroom. We sincerely hope, though, that you *will* consider breaking away from using formal instructor's guides as teaching tools, if that has been your practice. Give it a try—we feel the idea is important.

Lesson Planning: A Recipe for Success?

Maybe so; maybe not. Lesson planning, unfortunately, is not a "cookbook" subject. We simply cannot guarantee that if you mix these ingredients in these quantities, the result will suit everybody's taste. Like cooking, creating effective lesson plans is an art, and, in all honesty, not everybody is a master chef. All we can urge you to do, in creating both formal and personal lesson plans, is to give it your best. Experience will tell you what works and what does not, and experience will increase your proficiency in either developing formal lesson plans for your agency or personal lesson plans for your own use.

Lesson plans are important tools of our trade; they standardize teaching, they keep us on schedule and on course, they ensure that equipment and supplies will be ready and available when we need them, they keep us from leaving out essential teaching points, and they give us confidence when we step to the front of the classroom. Becoming proficient in their development and use is not a matter of choice; it is a necessity.

During our discussion of the presentation step, we briefly touched on methods of instruction. Like the ability to plan lessons effectively, developing the ability to teach using a variety of methods is not something you can put off or avoid. It is a fundamental part of our craft.

Let's go back to the classroom. It will be a good spot to talk about the different methods, and while we are there we'll give you some tips on what works and why it works.

6

Methods of Instruction

Each class session or lesson involves a body of information which the instructor must somehow convey to the students. Methods of instruction are the procedures or techniques we use to communicate or present the knowledge, skills, and attitudes our students must acquire in order to achieve the objectives stated for the lesson. Some methods are primarily verbal; others are visual. Some methods require students to participate physically; others do not. All of them, we hope, will actively involve the students mentally. No one method provides a "universal" teaching tool; each has advantages and disadvantages. An effective fire service instructor must know how to select and how to use the method appropriate for each segment of instruction.

Moving from the simplest to the most complex, we will consider the following methods of instruction:

- Lecture
- Illustrated lecture
- Guided discussion
- Demonstration
- Skills instruction

By arranging them in this order, we can build on lecture as a common foundation. Lecture requires the instructor to use only the voice. Illustration adds the use of a variety of two-dimensional visual materials. Guided discussion provides, for the first time, direct student involvement and feedback. Demonstration introduces all types of three-dimensional training aids for use by the instructor. Finally, students work "hands on" with the aids and equipment during skills instruction. Each succeeding method requires the instructor to become involved with an additional physical element or an additional technique. The common denominator, the one thread which is woven into the fabric of all the methods, is the voice of the instructor—the spoken word. As we will see, selection is seldom a matter of "either/or"; more often you can best accomplish your goals by adding and combining methods, mixing and matching the method used to the subject being taught or the kind of student involvement you want to achieve.

Lecture

Description: Speaking, telling verbally
Resources: Voice, gestures, teaching notes, recall
Advantages: Permits rapid presentation of information or facts
   Relatively easy to plan, organize, and rearrange
   No production costs, no equipment required
   Appropriate for large groups
Disadvantages: Limited student retention, single-sense involvement
   Does not permit active student participation
   Instructor cannot judge student understanding
   Can be dull and dry if not well performed

There Is Some Bad News . . .

Pure lecture is exceedingly difficult to do well. For a lecture to stand alone the lecturer must be a masterful speaker, the subject must be of vital interest to the students, and the information must be arranged and presented with great care—three difficult standards to meet for an afternoon class on salvage and overhaul. Since lecture relies heavily on one single student sense, hearing, student retention, without reinforcement by seeing or personal participation, tends to be low. Lecture is also the weakest method for transferring mechanical concepts and manual skills, areas which make up a large part of our training programs. For all these reasons, fire service instructor trainers tend to treat lecture as a necessary evil which should be used as little as possible.

This attitude leads instructors, especially new instructors, to rely too heavily on prepared audiovisual aids, particularly slides and films. We wind up with lessons overcrowded with media and instructors who spend too much time on selecting or developing aids and too little time on planning the logical development of their teaching points and lesson plans. We feel, very strongly, that *people* teach. Audiovisual aids can only "aid." They can help you to clarify or explain, they assist in reinforcing student retention, but they *cannot* and they *will not* substitute for logical words spoken with conviction by a motivated human instructor present in the classroom.

. . . And Some Good News

The lecture has the distinct advantage of being extremely easy to organize and, if necessary, reorganize. Since only your teaching notes are involved, and since there are no visuals or other aids to consider, you can make a change in sequence or emphasis quickly and easily. These same considerations point to the economy of the lecture method—production costs—other than your time, obviously—are virtually nil.

The lecture permits you to cover a great deal of material in a relatively brief span of time. It works well when you need to lay down a broad informational or motivational base quickly. The size of the group is immaterial, provided that they can hear. You can lecture to five people or five hundred (or five thousand, for that matter) with exactly the same investment in time and effort.

We seldom teach for any extended segment of instruction using pure lecture only; more often lecture will be used to introduce new topics or concepts, to argue for or against a particular idea, to make a transition between segments taught using other methods, or to summarize. Where lecture stops and some other method starts is not important. What *is* important is the understanding that lecture, the unvarnished spoken word, *does* have an essential place in your kit of teaching tools. Verbal communication is the common thread which runs though and ties together all of the other methods of instruction.

Getting Involved

The lecture portions of your classes should be relaxed and relatively informal. Feel free to move away from the lectern, to move among the students as you speak. Our constant challenge is to build and maintain the bridge of communication with our students. There is no better way to get this job done than by talking earnestly, eye to eye, with the class. Don't be

Fig. 6-1. Get excited!

afraid to show some emotion. If you feel strongly about a point, let your students know it (Fig. 6-1). Getting the students totally involved in the subject should be our goal in every class. How can they become enthusiastic if you don't let your own excitement show?

Telling Isn't Teaching . . .

One final word of caution—lecture is a one-way street. Information flows from the instructor to the students with no feedback to ensure understanding. Remember, just because students seem attentive, perhaps even nodding in apparent agreement, is no guarantee that they are mentally following you. Students, adult students in particular, will drift off into their own thoughts, leaving you feeling that you have done a superb job of teaching when in fact they have not absorbed one single idea. We said it before, but it is worth saying again: "Telling isn't teaching; listening isn't learning." It may be worthwhile to print those words on a card and tape it inside the top of your briefcase.

Illustrated Lecture

Description: Showing with verbal commentary

Fig. 6-2. "One picture is worth . . . "

Resources: Voice plus graphics and projections (any two-dimensional displays)

Advantages: Provides visual images
Increases student interest
Enhances retention—multi-sense involvement
Improves understanding
Discloses otherwise hidden mechanical operations
Provides wide access to places, processes, and equipment through slides, film, and video

Disadvantages: Entertainment aspect frequently overshadows teaching value
Requires substantial planning and skill to integrate visuals smoothly into lesson flow
Most visuals are expensive to purchase or time-consuming to produce
Projections require special equipment
Some visuals are too small or poorly designed

The illustrated lecture, or illustration method, adds visual images to the instructor's words. The sense of sight provides instant, uniform communication which cannot be achieved by words alone. If we verbally describe a house as: "two-story white stucco on a corner lot, with black shutters, a gray shingle roof, and a front porch with a wrought-iron railing," each person hearing the words forms a different mental image. One picture would communicate the scene immediately and uniformly; everyone will see exactly the same thing and be far more likely to retain and be able to recall the information (Fig. 6-2). Illustration, as a teaching method, involves the use of all graphics and projections: photographs, drawings, charts, slides, overhead transparencies, motion pictures, videotape—any material which can be drawn, printed, or projected on a two-dimensional surface.

While illustrated lecture swings our attention to the visual aspect of teaching, don't overlook the importance of the words which accompany the visuals. Regardless of the quality of the visual images, the commentary is the key element. Visuals can make a good lesson better; they will *not* make a bad lesson good.

Quality and Design

As instructors, we face a constant need to focus the attention of each individual student on the subject we are discussing. Visual aids provide a means to accomplish this by directing the eyes and channeling the thoughts of our students to the points which are significant. They also assist in holding the interest of the class by helping to shut out unrelated casual thoughts. To do this effectively, our visuals must be of first-rate quality. They should be simple, and they should contain no irrelevant material. They should be attractive, with color arranged carefully to emphasize main points or parts. Students will judge what we show them by comparing our materials, consciously or subconsciously, with extremely expensive advertising media (Fig. 6-3). There is no way we can compete, dollar for dollar, but we can, at least, take care not to use aids of poor or questionable quality. Be particularly watchful for incorrect spellings, incorrect grammar, and lists containing inconsistent elements or inconsistent syntax. Most certainly, all visual aids must be absolutely accurate; all principles, facts, and figures complete, correct, and current.

It should go without saying that all visual aid items must be big enough to be visible, but experience suggests we should say it just the same. Too many times instructors introduce visual material by saying, "Now, I know you can't see this but . . ." (Fig. 6-4). What possible value can be found in using a visual aid which students cannot see? Check personally before using an aid; view it from the perspective of the student group. The smallest significant detail must be large enough to be seen by the most distant student.

To the greatest extent possible, each visual panel should introduce a single point. Presenting one thought or concept at a time is less distracting, and it focuses student attention exactly where we want it. This concept is illustrated in those slide sequences which sharply highlight new words or phrases while repeating previous words or phrases in low-key color or smaller letters.

For our own benefit, visual aids should be portable, durable, and

Fig. 6-3. Advertising sets the standard.

manageable. They should be easy to transport, assemble, and disassemble. Teaching is a tough job. It is physically and emotionally draining. You should not have the additional burden of fighting a bulky, unwieldy, poorly designed visual aid package. Visuals should be able to withstand a bit of rough handling and repeated use. A little extra expense at the start, to build in better durability or to provide protection during transport, will save time and money in the long run.

Aids should be practical, easy to operate and manipulate. Visual aids should illustrate your message without breaking the continuity of the lesson. They should be designed and arranged so they can be presented with a minimum of noise, motion, delay, and other distractions. Student attention should be riveted to the message, not to the equipment or the technique being used to produce the message.

Standardization of Instruction

A well-designed visual aid package will promote standardized instruction. The teaching of an individual instructor will vary less from one class to the next, through the use of a selected set of visual aids. In organizations where different instructors teach the same material to different groups, standardized visuals will encourage uniform teaching of the desired lesson objectives.

A Word to the Wise . . .

Strange as it may sound, visual aids do not necessarily make teaching easier; they often make the instructor's job more difficult. Harmonizing multiple teaching aids into the lesson continuity is not easy.

Be sure any aid you select truly supports the instruction you plan. Don't

Fig. 6-4. "Now, I know you can't see this, but . . . "

be too quick to alter your lesson plan to incorporate visual material "just because it is there." Instructors too often fall into the trap of using films, slides, and other aids just because they are readily available, even though they have little value in terms of what is to be taught in that lesson. The only valid question you should ask is: will this visual aid help my students to learn more effectively that which they are supposed to learn? If the answer is "Yes," use it. If the answer is "No," discard it.

Guided Discussion

Description: Telling verbally and/or showing while obtaining student involvement through questions

Resources: Voice, notes, recall, and any appropriate visual material plus prepared and offhand questions

Advantages: Provides feedback; ensures understanding
    Establishes student involvement
    Promotes individual thought
    Increases interest
    Improves retention

Disadvantages: Difficult to keep on planned schedule; time-consuming

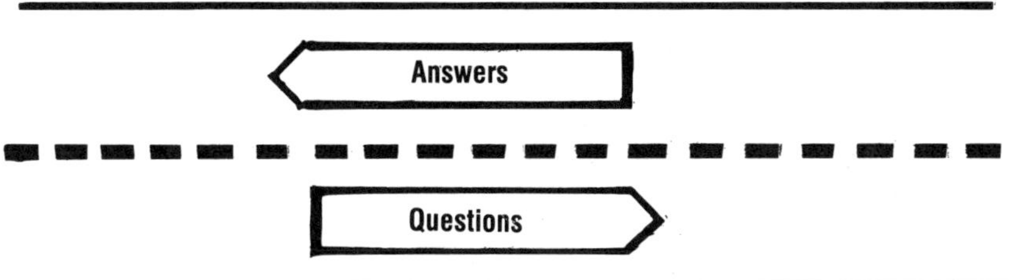

Fig. 6-5. Guided discussion—the two-way street.

Difficult to control direction of discussion; easy to drift away from lesson objectives
Requires careful and time-consuming planning
Requires substantial practice to perform with ease and skill
Requires substantial proficiency in the subject area under discussion

Student Involvement at a Price

Guided discussion gives us an important element not found in lecture or illustrated lecture: direct student involvement in the lesson. Personal participation will make a class session much more interesting to your students, and the accuracy of their comments supplies you with immediate feedback on how well they are grasping the information being presented (Fig. 6-5). As always, nothing is free. In return for the benefits obtained, we pay a substantial price in both class time and preparation time. The result is well worth the cost. There is no other way to achieve the vital two-way contact which guided discussion provides.

Guided discussion is usually initiated and sustained by the instructor. We can also be pulled into this method of teaching by our students when we had something else planned. When a student interrupts with a question during class, you have several options: *(1)* to answer the question at that time and then return to your planned lesson; *(2)* to ask that the question be held, and defer answering until some later specified time; or *(3)* to use the question as a springboard to dive into guided discussion. The relative pertinence and the timeliness of the question are important considerations. The path you choose will depend on your assessment of what will benefit the class most at that moment.

Control and Direction

The word "guided" is paramount. Effective use of guided discussion requires control and direction. You must walk a fine line between dominating the conversation, on the one hand, and totally losing control, on the other. A rambling "bull session" may be fun and may even be informative, but it is not teaching and has no place in a structured class situation. A good instructor, teaching a well-organized course or lesson, will rarely have enough time to permit the students to run away with a discussion.

Control is achieved by knowing what you want to accomplish and by having built into the lesson plan prepared questions which will move discussion in the direction you have selected. This method has the added benefit of forcing you to identify the really important points and significant concepts which the students must comprehend. Setting up a guided discussion is much more complex than merely dropping a few questions into a canned lecture. You need to plan carefully the thought processes you want the class to follow, and you must determine how much time you can afford to spend. In addition to the prepared questions of the lesson plan, you will have to think on your feet to develop the offhand questions necessary to keep the discussion moving according to plan.

Taken as a whole, guided discussion is not an easy job, but one which is absolutely essential. An instructor who does not become a master of questioning technique, a master of posing questions and competently handling student responses, a master of gently but firmly guiding discussion will not be able to teach others effectively.

Types of Questions

There are two basic types of questions: direct and overhead. A direct question is specifically directed to an individual student, by name. An overhead question is asked of the entire class and left "hanging in the air" for anyone to answer.

Direct:

"What is the weight of one gallon of water—Jim?"

Overhead:

"What is the pressure, in pounds per square inch, at the bottom of a column of water one foot high?"

Direct questions are asked by using a technique known as "ask, pause, call." Ask the question first, pause for a few seconds, and then call the selected student by name. The rationale behind "ask, pause, call" is the suggestion that, until you identify a specific student, the entire class must think about an answer to the question. If you name a specific student first, so the theory goes, the rest of the class tends to relax mentally. We're not convinced this idea is a vitally important point, but we present it for your consideration.

The only other type of question worth mentioning is relay. The concept of a relay question is that the instructor, instead of responding, relays a student's question (or a partial or incorrect student answer) to another member of the class. This device is obviously a useful technique, but it comes about so naturally that we doubt it needs to be introduced so formally. We must, however, note one particular point. If you do not know the answer to a question raised by a student, say so, and *then* relay the question to the entire class to see if anyone can answer it. If the question is relevant, and if it is not satisfactorily resolved, tell the class you will find the correct answer and pass it on to them at a later session. Do not, under any circumstances, try to bluff your way through. Sooner or later, your bluff will be called, and your credibility will be totally destroyed.

Questioning Techniques

The manner in which we use questions can work for us or against us. Avoid directing questions to students by using any readily recognizable pattern—alphabetical order, for example. Human nature being what it is, that *will* cause students to relax, knowing that once you have passed them they are "off the hook" until the next round. Call on students in all parts of the room, and be careful not to fall into the trap of directing questions only to the better or quicker students. With a little care in matching the level of question difficulty with the level of student proficiency, you can involve the entire class in the discussion. It may be particularly useful to single out any inattentive or unruly students for a little special attention to help pull them into the mainstream of activity. Guided discussion is intended, to the greatest possible degree, to involve the entire class.

Be sure everyone can hear all questions and responses, yours and those of other students. With large groups, it is often necessary to repeat student questions or answers to ensure that people in all parts of the room can follow what is going on. Be careful to avoid becoming involved in a lengthy, two-way conversation with any one particular individual; it is discourteous, and most groups resent one member dominating any significant amount of class time.

An effective instructor will ensure that everyone who wants to contribute has an opportunity to do so. You must direct the verbal traffic, from time to time, identifying who may speak next and who must wait. It is a good idea

Fig. 6-6. "And who is buried in Grant's tomb?"

to comment favorably on as many student answers as you honestly can. Praise encourages participation.

Question Preparation

Invest some time and thought in preparing and arranging the questions you plan to use, particularly those written into your lesson plan. They should be perfectly clear, carefully worded to avoid any ambiguity, and relatively brief. In general, your questions should require students to think out and frame an answer, not to answer with just a yes or no. Catch or trick questions should usually be avoided, particularly if they cause student embarrassment. If you routinely inject this type of sneak attack into the questions you pose, students will, quite logically, become reluctant to volunteer answers. Finally, try to avoid including the answer in the body of the question (Fig. 6-6). All of us, from time to time, say something truly stupid such as "What color was General Washington's white horse?" Not much you can do but laugh it off an try again.

Fine Tuning

After you have taught a particular lesson three or four times, you will develop a feeling for the type and scope of answers a particular question will generate and you can, perhaps, fine tune the wording to accomplish exactly what you want.

Questions as Teaching Tools

Guided discussion emphasizes the use of class questions to stimulate discussion, to guide student thought, and to direct student attention. We can, at the same time, informally evaluate both student progress and our own effectiveness. Questions can also be useful in settings other than guided discussion. At an introductory class session, questions can help us to determine what knowledge students bring with them, what abilities and interests they possess which will be useful in the work we are beginning together. During class activities, projects or skills practice sessions, for example, questions give you an excellent means of providing low-key guidance to students in planning their work or analyzing problems.

Finally, questions are an effective tool to use in review or summary, giving us an opportunity to touch on and reemphasize the essential points of the lesson, while at the same time checking class comprehension of the material covered.

Keseping One Step Ahead

Guided discussion in particular, and questions in general, force us, the instructors, to keep up to date with technical developments in the fire service and to stay mentally wide-awake in class. Compared with lecture or illustrated lecture, guided discussion is much more demanding in requiring us to have both extensive personal knowledge and an ability to think on our feet quickly and precisely. In lecture and illustrated lecture the instructor is in total control, and students are permitted only to receive. Questions invite class members into verbal combat, and fire service service people are sharp. For a good instructor, trading wit and wisdom with a motivated class is the greatest fun that teaching has to offer, but you had better have your act together. And keep one eye on the clock—guided discussion will use up a fifty-minute class session faster than you ever dreamed possible.

Demonstration

Description: Doing, with verbal commentary
Resources: Voice plus equipment, mock-ups, models, any three-dimensional
  aids

Advantages: Provides absolute proof that the equipment or process being demonstrated really works
Enhances interest
Clearly shows correct procedures which provide the basis for skills instruction
Increases motivation
Disadvantages: Requires thoughtful and careful preparation to ensure performance is correct and complete
Time-consuming to prepare, practice, and perform
Students may miss the teaching points unless these are made clearly and are not lost in the demonstration activity
Visibility may be a problem if the training aid is small or the group is large
Equipment is required; costs can be high

Similar but Different

Demonstration transmits information to students through sight and hearing and is similar, in that respect, to illustrated lecture. Demonstration is different, however, from any of the previously discussed methods because it requires us to work "hands on" with the tools of the trade while simultaneously describing the main steps and key points of the procedure or the process being demonstrated. Objects used include all types of actual equipment, cutaways, models, mock-ups—any three-dimensional materials which will increase or reinforce student understanding.

Practice, Practice, Practice

Demonstration places some special and unique burdens on us in addition to requiring the ability to work skillfully and talk intelligently at the same time. If you are teaching the subject for the first time, or have not performed the demonstration recently, practice is absolutely essential. You must, repeat *must*, be able to do the work involved, and do it flawlessly, without thinking about what you are doing. The reason, clearly, is that your mind will be occupied with describing the process, ensuring that you touch on all the main steps and key points, and, perhaps, responding to student questions. We do not believe it is possible to practice too much. Think about the top-flight athlete who, no matter how good or how experienced, is constantly practicing in order to hold or improve that essential competitive edge. Like the athlete, you will be watched closely and critically by your audience. Be sure your performance merits the time and attention your students are investing.

Errors and Omissions

In spite of our best efforts and intentions, mistakes will happen. If you make an error or leave out some significant step during a demonstration, stop immediately. Explain fully and clearly what went wrong and why it went wrong. When you have explained the problem and answered any student questions the situation may generate, back up to some convenient point in the procedure and try again. If it is necessary, describe where you are picking up the demonstration and explain why you chose to re-start at that point so the students understand exactly what you are doing.

Physical Preparation

All equipment must be in place and properly arranged before class. For subjects you teach regularly, a checklist of the items required will be a useful addition to your lesson plan folder. Without a checklist you will, sooner or later, forget some essential item. Arrange your aids for minimum class disturbance and maximum instructor convenience; have all needed items close at hand and arranged in the order of use. When you must utilize large pieces of equipment, it is usually more desirable to move the student group to the aid rather than the other way around. Such moves should be made with clear instructions to the class so students do exactly what you want with minimum delay. If it is at all possible, make the transition to the new location during a break period to minimize wasted time.

If the equipment being demonstrated has one or more moving parts, Mr. Murphy's well-known law becomes operative: "If something can go wrong it will, and always at the least convenient time." Check every piece of equipment carefully before class to ensure that all parts are present and cooperating. When you are sure everything is ready, check again.

Keep an Eye on Your Group

In any demonstration, visibility is critical and much more difficult to achieve consistently than in illustration. Many demonstrations, probably *most* fire training demonstrations, require that the student group cluster around the instructor and the aid. The instructor must keep an eye on the group arrangement and make sure that all students have a clear line of sight. The position you assume while working can often block the view of a portion of the class (Fig. 6-7). It may be advisable or necessary, at times, to do at least part of a demonstration from an unconventional position—working from behind a small piece of equipment, for example, reaching over the top to

Fig. 6-7. If they can't see, they can't learn!

manipulate the device, so students can see. If correct positioning of the operator is critical to student understanding, repeat the process in the proper position, several times if necessary, until you are sure every student has had an adequate opportunity to observe.

After the initial demonstration, instructors often become involved (improperly) in a question-and-answer exchange with the students nearest to the training aid. This situation can easily drift into a private teaching session for those at the front of the group. Students who are not involved, or, worse still, cannot hear or see what is going on, will quickly lose interest. Check frequently to make sure you have the attention of the entire group. Maintaining eye contact is difficult during most demonstrations, but that fact is no

Fig. 6-8. When noise is a problem: Teach   demonstrate   review.

excuse for not trying. As always, be sure the entire class hears all questions and all answers. Your summary should review the entire procedure and ensure that students understand all main steps and key points.

If They Can't Hear, They Can't Learn

With some fire equipment demonstrations, working at the pump panel or with gasoline-powered tools, for example, noise is a frustrating problem for both the instructor and the students. Consider following a procedure of describing first what you are going to do, identifying the main steps and the key points, indicating significant controls and operations. Then start the equipment and perform the demonstration. Finally, shut down, review what was done, and call for questions (Fig. 6-8). Trying to teach against the roar of a motor is a waste of time for all concerned. You will find yourself shouting and the students, unless they can read lips, gain little or nothing.

Skills Instruction

Skills instruction is not a "method" in the same sense that lecture, illustrated lecture, guided discussion, and demonstration are identifiable and

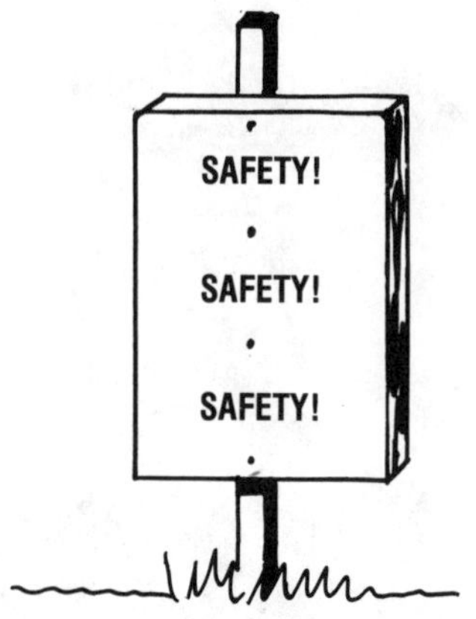

Fig. 6-9. Always put safety first.

fundamentally different teaching techniques, each with certain advantages and disadvantages. The teaching of skills is, however, a unique and essential part of fire service instruction. Many current texts and instructor training programs treat skills instruction as a subset of demonstration. We disagree. While demonstration plays an important part in teaching skills, it provides only the first step. There is much more to the process.

In demonstration, the instructor is working with the equipment or procedure while students watch and listen. Demonstration gives the student an opportunity to observe the task and to think about personally performing the motions involved. Both are essential steps in preparing the student for doing, but *learning* cannot truly occur until the student takes hold of the tools or equipment and attempts to duplicate the action. Crucial to student success is the way in which the instructor bridges the gap between "observing" and "doing." Crucial, also, is our orchestration of student progress through these first tentative attempts to perform any new task.

A Word of Warning

Remember, skills instruction is the activity where students can be hurt. At every stage of your skills instruction activities—planning, execution, and critique—make safety an active and visible part of the process (Fig. 6-9). In training for dangerous tasks, people may be injured in spite of our best efforts. Just make sure any injury to *your* students is not the result of *your* failure to plan and carry out every appropriate safety measure properly. In particular, during hazardous instruction, do not permit any unplanned,

TEACHING STEP	INSTRUCTOR PROVIDES	STUDENTS LEARN
Show Them How Tell Them How Let Them Do It	Main Steps Key Points Practice	What They Must Do What They Must Know Proficiency

Fig. 6-10. The learning sequence.

"Let's try it just once," experimental activities. Scuba divers have an excellent slogan, "Plan your dive and dive your plan." Stick to *your* plan.

Show and Tell

As we said in Chapter 4, skills instruction has been, for many years, taught by the time-tested technique known as "show and tell." "Show 'em how, tell 'em how, let 'em do it." It has been employed for generations, and it still works today—very simple, but most effective.

Today we use terms like "psychomotor skills" and we have, without question, a much better understanding of the entire learning process. The basics, however, have not changed. Skills instruction is still a combination of showing how a task is done, telling how a task is done, leading students through a slow and distinct series of performance steps, and, finally, providing the opportunity to practice under guidance (Fig. 6-10). The blending and timing of these stages, obviously, will be affected by the complexity of the task or equipment under study and by the knowledge and experience of the students and of the instructor.

First Things First

In skills instruction, some of the matters we touched on earlier suddenly become extremely important. Remember the difference between "seeing" and "perceiving." Students may clearly *see* your demonstration and yet not *perceive* some essential point or points necessary for proper performance. Be patient.

The law of primacy plays a strong part. Carefully plan the order in which you present the various elements of a skills teaching session. It is essential that the flow of information be logical, easy to follow, and absolutely correct.

Main steps and key points should be presented in proper order and precisely identified. Students must, at all times, be able to understand clearly and unmistakably what you are doing, why it is being done, and what you expect them to do next. When they begin actual practice, early success and frequent success is necessary to reinforce student learning and student confidence. Build it into your lesson.

Most skills force students to acquire a sense of the kinesthesia involved. It can be terribly difficult for both the instructor and the student when this perception of the motions and forces involved fails to become apparent. Again, be patient, and in all skills areas be alert for ways to make your job, and the student's job, easier. Sometimes a "gimmick," a catch phrase, or an analogy can produce understanding where mere repetition fails. Rabbit holes and trees may sound a little silly, but they have helped a lot of klutzy people learn to tie the bowline.

Mirror Image

Remember that a skill demonstration conducted facing the class provides the students with an entirely different view of the action from the one they will see "hands on." Knots provide a good example. With the instructor facing the class, the student's view of the instructor's hands tying the knot is similar to looking into a mirror—not exactly the same, of course, because of the left hand/right hand reversal, but similar (Fig. 6-11). If this positioning is an important consideration in the skill being taught, turn your back to the class, hold your hands high or in some other clearly visible position, and show the action in the proper plane and perspective from the viewpoint of your students.

Remember, also, that right- and left-handed students may need to work differently. Individuals should be allowed the greatest possible freedom to perform a task in the manner most comfortable and effective for them, provided that manner is consistent with safety and meets the student performance objectives.

Whole-Parts-Whole

Remember whole-parts-whole? First, describe the entire process while you perform the action. This initial demonstration should probably be done at normal working speed to show how the completed task will be performed in an operational setting. Immediately following this first showing, if "normal working speed" is too fast for students to follow the action, or if the task

Fig. 6-11. Skill demonstration—mirror image.

is complex, repeat the action slowly, identifying the main steps as they are accomplished and discussing the key points as necessary.

Too often, instructors are tempted at this point to "show off" their personal proficiency. Repeated demonstrations at expert speed almost always fail to improve student understanding and may, in fact, convince students that the task is totally beyond their capability. Resist the temptation to dazzle the class with your skill. They expect you to be good. It should not be necessary to prove it, particularly in front of a class of neophytes who are already convinced you can leap tall buildings in a single bound. It may be fun but, unfortunately, it isn't teaching.

Step by Step

Following the "whole" demonstration, proceed to the "parts." Slowly and carefully lead the students through each sequential step (Fig. 6-12). Ideally, each student should have the necessary equipment so the entire

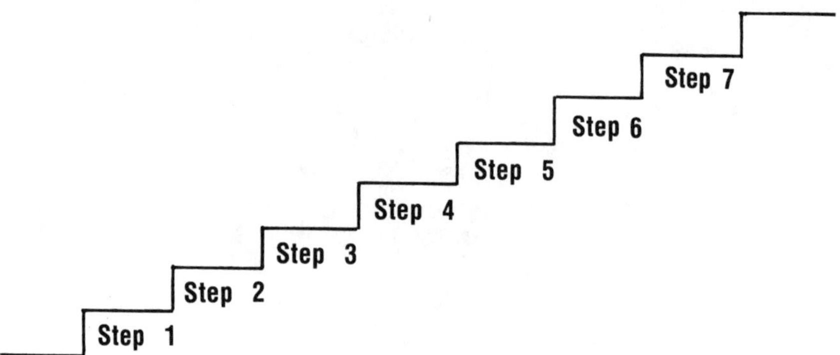

Fig. 6-12. Slowly and carefully, one step at a time.

class can follow together. If that arrangement is not possible, the steps must be repeated as many times as necessary to provide practice for everyone.

If the entire class is following step by step, you have to insist that they stay together, at least for the first "walk through." At each step, check carefully to ensure that everyone is ready to move on before proceeding to the next element. The process, fundamentally, is: instructor shows and tells—students do—instructor checks. Holding back the quicker students usually is not a problem; judging how long you can hold up the entire group while waiting for one or two slow students to catch up is more difficult. It may be necessary to proceed even though one or two are left behind. Take time to assure the slow learners that you know they have a problem and that you will get back to them to provide individual help. You can often "pair" quick and slow learners. Use the more proficient person to assist the one having difficulty. This method may help the slower student to keep up with the class, and it will also maintain the interest and reinforce the learning of the better student.

Once the whole group has been through the entire sequence, you can apply several different techniques to reinforce the process; instructor tells/students do, or students tell/students do, or simply go on to a free practice session in which each student works at his or her own pace while you circulate, monitor progress, and assist as necessary.

Developing Proficiency

When students are ready to perform "whole," practice should proceed through the three levels of increasing proficiency we mentioned back in

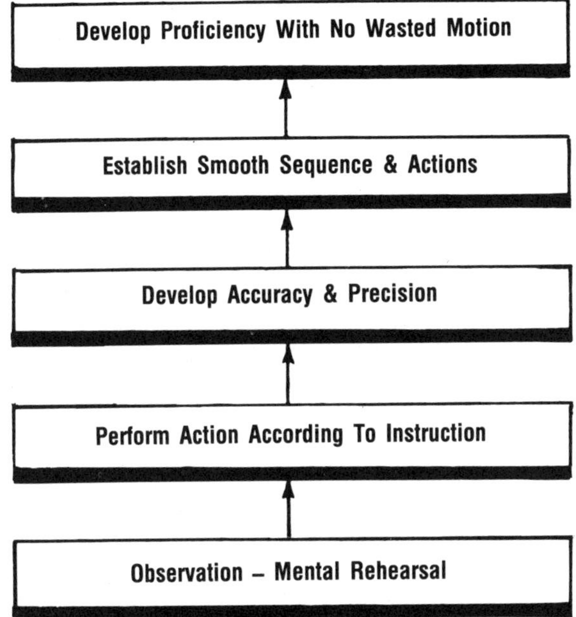

Fig. 6-13. Developing skills proficiency.

Chapter 4. How long this takes, how many practice sessions will be required is, of course, a function of the difficulty of the task and the adaptability of the student group. Students should first become able to perform all steps correctly and in the proper sequence without instruction. Once this stage is mastered, students should develop the flow of the action, smoothing the transitions from one step to the next into a coordinated whole. Finally, students should concentrate on eliminating all unnecessary motion, and strive for the ability to perform at a high level of proficiency (Fig. 6-13). In the process of developing most physical skills, speed comes with sureness. Pushing for speed before proficiency has been developed will result only in inaccurate or slipshod performance. Don't let students practice too fast, too soon. Slow and right is much better than fast and wrong, especially in the beginning.

Simple to Complex

If the entire skill area under consideration is extremely lengthy or complex, or is made up of a number of elements, it may be necessary to teach a series of separate "lead up" sessions before putting the complete operation together.

Interior structural firefighting, as we said, requires a variety of skills which cannot possibly be taught simultaneously. Students must first become completely proficient in a number of relatively simple skills: donning of protective clothing, handling hose, operating nozzles, wearing and working in breathing apparatus, and, most probably, demonstrating an understanding of the basics of fire behavior (Fig. 6-14). Only after mastering an entire series of simpler skills is the student properly prepared and ready to attempt a complex skill evolution like interior firefighting. This technique of moving from simple to complex is applicable in many areas of fire service training.

Alternative Approaches

If a skills session group is large and instructional staff is limited, it is possible, under certain circumstances, to use students as assistants.

If students present are known by the instructor to have mastered correct skills through previous courses or other advanced work, they can be assigned to act as squad or section leaders and can assist or monitor the progress of others.

Another useful technique is pairing students and having them alternate in the role of observer and learner. The observing partner should be instructed to watch for correct procedures, correct sequence, and safe practices.

Instructor Tips

One of the most difficult things for a new supervisor to learn is the absolute necessity of *letting* a subordinate do a task which the supervisor can do better. It is very difficult not to say, "Move over; let me do it." So it is with instructors and students. You must resist the temptation to take over the tools and do it yourself, justifying your move by having them *watch* you one more time. Sometimes, of course, that action may be necessary, but sooner or later you must move back and let them struggle.

If a student is having difficulty, use questions to try and solve the problem. Has a step been left out? Is the equipment adjusted properly? Are the person's hands in the proper positions? Questions, skillfully used, can direct the student's attention to the proper solution without your active interference with the work. Attempt to use a positive approach; tell them what *to do* rather than what *not to do*.

Finally, ask questions or offer further instructions only if a student is

Fig. 6-14. Master the simple before attempting the complex.

having a problem. If students are making progress, give them an "Attaboy" and move on to somebody who needs your help.

Selecting the Proper Method

In Chapter 3 we talked about teaching knowledge and skills and changing attitudes, putting each into a neat little box. In actual practice, it is seldom possible to treat or teach them entirely separately. In the fire service, we rarely teach knowledge without relating it to some practical application. Also, we rarely teach skills without first establishing some form of knowledge base. Attitudes are often interwoven into both knowledge and skills instruction. Because the types of information to be transferred overlap, methods of instruction overlap. Rarely do we select and use only a single method, even

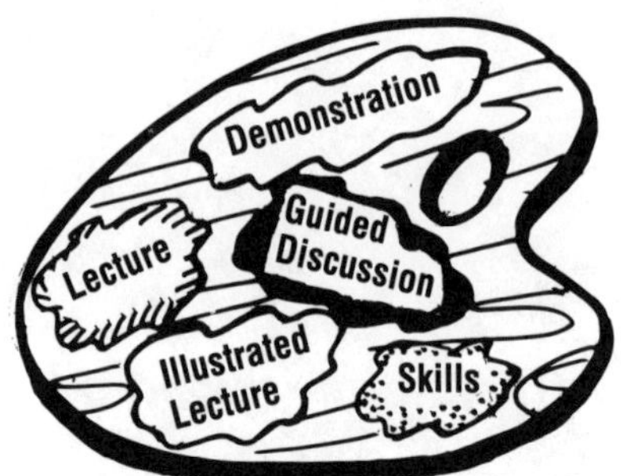

Fig. 6-15. Blend them skilfully.

for a simple fifty-minute lesson. More often you will find yourself using a combination of methods, moving back and forth among them as you progress from one teaching point to another.

Selection Criteria

The method of instruction must fit the kind of information being transferred. Using the wrong method can be like trying to carry water in a sieve; no matter how hard you work, you are not going to move much material. Common sense tells us we can carry ice cubes in a sieve but not water. Common sense tells us we can teach facts by lecture but not physical skills. There is no magic in selecting the "right" method of instruction, just common sense. Be sure the method you select is appropriate for the information you want to convey.

Other factors can also influence your selection. Instructor preference must be mentioned even though we should strive to become proficient in all methods. Each of us has different experience; each of us has a different personality. We have different abilities with the spoken word, different abilities to letter or draw or spell. It is logical, then, that we may be more comfortable with a particular method or, at least, a particular medium.

The equipment available can also influence selection. The type and quantity of training aids available for use by the instructor and the type and quantity of tools and gear available for use by the students can affect your decision.

Finally, the knowledge base of the students is an important factor. Teaching recruits will heavily involve lecture, illustrated lecture, and skills instruction. With new personnel, guided discussion may be restricted to periods of summary or review. With an experienced group, on the other hand, such as a tactics class, guided discussion may be the primary method used.

Common sense will point you in the right direction. Experience will either confirm your decisions or suggest changes which should be made. The methods available to us are like the range of colors on an artist's palette. Properly blended and skillfully applied, they can be used to create a masterpiece (Fig. 6-15). In every class we teach, that should be our goal.

We probably need to stay right here and, next, talk about ways to get our students actively involved in the learning process. These chairs are starting to feel hard, though, and there is an old fire service teaching proverb which says: "The mind can only absorb what the behind can endure." Let's take a little stretch break and then we'll push on.

7

Special Application Techniques

Remember the four steps: introduction, presentation, application, and testing? If you stop and think about it, the methods of instruction we have just discussed are used mostly in the "presentation" step. They are the methods by which we *present* new knowledge, new skills, new values, conveying information from our heads and hands and hearts to the heads and hands and hearts of our students. Following this instruction, following the presentation step, we must involve our students in some form of activity in which they can personally *apply* the information they are learning.

In some areas, skills instruction, for example, the line between presentation and application gets a little fuzzy, but it is still there. Actually, the absolute definition of any line between them is critical only in lesson preparation, making sure that we have covered both steps adequately in our teaching plan. In the classroom or on the drill ground, teaching occurs during both presentation and application. The major points of instruction will, obviously, be given during presentation, but you will continue to teach your students during the application step—coaching them, reminding them, and correcting them, as you "fine tune" their abilities. Your teaching increments become smaller and smaller as the abilities of your students increase, but you, like the boot camp drill instructor, will continue to apply that last-minute spit and polish right up to the moment they walk out the door at the

Fig. 7-1. Buzz group.

end of the final class. Let's take a look at some techniques for involving students in the application of the principles and practices we have taught.

Conference

We did not mention "conference" under methods of instruction although the term appears in many instructor training courses. We left it out because, while we have frequently seen it *taught*, we have never seen it *used* as a method of instruction. True, there are many conferences held each year by fire service organizations, but these meetings are usually administrative or creative in nature, *not* instructional. One type of "mini-conference," however, *is* frequently used in teaching; it is known as the "buzz group."

Buzz Groups

The name "buzz group" comes from the buzz of conversation which this technique causes (Fig. 7-1). Buzz groups are most commonly used to provide an application step in cognitive or knowledge subject areas. After teaching some particular method of problem solving, the instructor divides the class into groups, usually of four to six persons. These numbers will form groups large enough to provide the essential "give-and-take," yet small enough to pull everyone into the group discussion. It is necessary that the members of each group cluster together, preferably around a table, so they can communicate and work together within the group and offer minimum interference to others working in the same room.

Each group is assigned a problem, and a time interval is set for group discussion and preparation of a group report. There is no particular established length of time for buzz group activities; most are in the 15- to 30-minute range, and they are rarely as long as an hour. Groups may be assigned the same problem and their solutions compared, or the groups can be assigned different problems in order to demonstrate a wider range of applications.

Group reports are presented by a member appointed by the instructor, or, more commonly, selected by the group. After the "official" group report is given, other members of the group are usually invited by the instructor to expand on, or to *dissent* from, the group report. The discussion is then opened to all members of the class to question, confirm, or criticize the group solution.

A common variation on this technique is first to present a problem or problems to the entire class and provide time for individual study and decisions. Members then enter the group with preconceived ideas of what they feel should be done and must hammer out compromise solutions among all those suggested.

The conference aspect of this learning technique lies in the joint effort within each group and in the class discussion, encouraged by the instructor, of the solutions offered. As in all conference leadership situations, the instructor takes a low-key role as a moderator or facilitator. Conference is usually considered to be a "self-teaching" situation in which the student group, within itself, seeks to create a consensus opinion of how a problem should be solved. You, as the moderator/instructor, can sneak in a little teaching within a buzz group session by circulating among the working groups to keep them "on track," by assisting students in defining or describing points of view, and by commenting on the solutions presented. Typically, the problems used in this kind of setting do not have "right" or "wrong" answers. Each solution offered is judged against the concepts which were previously taught in class. If the solution is consistent with those concepts, and the solution can be logically defended, it should be considered acceptable.

Case Studies

Case studies are knowledge application tools which can be used within buzz groups or as individual assignments. The characteristic element of a case study is reality. Case studies are selected from real-life tactical or management situations (Fig. 7-2). Students are asked to analyze the problem, to critique the actions of the actual participants, and to offer suggestions for improving upon the real-life actions described.

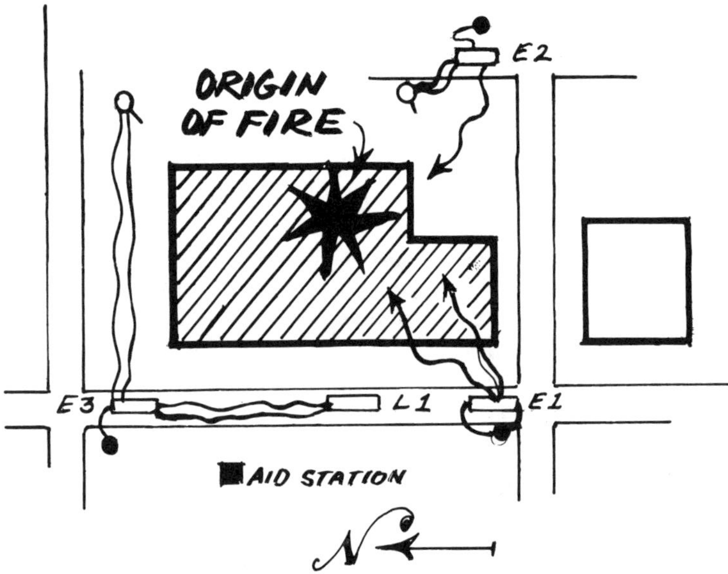

Fig. 7-2. Real situations provide case studies.

Again, the *logic* of the proposed solution is the key factor in judging student performance. Be prepared not to expect dogmatic right or wrong answers, and look instead for answers compatible with the principles the case study is expected to illustrate.

Projects

One step beyond case studies are application projects which directly involve students in a real-world setting. One of the most common fire service project assignments is having students conduct and report on a building or facility inspection. The technique is a useful one and it should be somewhere in your personal bag of teaching tricks. The inspection can be slanted in a number of ways, depending upon the course of study and on what specifically you wish to accomplish.

For a course in tactics and strategy, or fireground management, if you prefer, the assignment might be to conduct a pre-fire planning inspection, to record the information obtained, and to construct a pre-fire plan. In many classes, the pre-fire plan is then tested through some form of fire simulation.

An alternate purpose for an inspection assignment might be fire prevention: looking for heat hazards and fuel hazards and reporting on any potentially dangerous situation. Perhaps the emphasis might be on personnel safety—reviewing compliance with corridor, stair, and exit requirements and other elements of the Life Safety Code. Still another purpose might be to evaluate the "built-in" fire defenses of a plant or facility, checking against code requirements to determine the adequacy of extinguisher coverage, interior standpipe hose line spacing, sprinkler spacing and pipe sizes, and reviewing any other existing fixed fire protection devices.

If you assign an inspection project as a part of a class you are teaching, there are two areas which require your close attention. Internally, there must be a clear-cut understanding between you and the students as to what you want done, how their reports are to be presented, and what criteria will be used for grading. The time you put into preparing reasonably detailed instructions will absolutely be worthwhile, reducing the number of problems and complaints you will have later. Be specific in explaining the parameters of the inspection process and the format you will require for the reports they submit. Obviously, the depth of information you demand will be a function of the level of the class and the abilities of your students.

Working with Property Owners

Outside the classroom, you must see that property owners are not overly burdened by your class assignments, particularly if you are teaching in a jurisdiction where official inspections are a regular part of the fire department routine. Many business people will be happy to cooperate if they understand clearly what you are trying to accomplish; others will not. In assigning projects, or approving student project selections, try to make sure you are involving owners and managers who are willingly cooperative. Students must conduct themselves in a totally professional manner at all times (Fig. 7-3). Their assignments must not bring discredit on the local department nor interfere in any way with the normal flow of departmental business.

There are many types of projects other than inspections. You are limited only by your imagination. Be creative.

"Borrowing"

If the same assignments are used in several different classes, be prepared for students turning in work they have "borrowed." We have seen two "different" student maps so carefully traced that when one is placed over the other and both are held up to the light, they look as if they had been printed on the same press. Another common "giveaway" is discovering exactly the

Fig. 7-3. Cooperate with owners and managers.

same misspelled words in two or more "independent" reports. Such cheating is not difficult to catch, but it can be difficult to deal with if you are not expecting it. Be aware that your students *will* try to "con" you, if they think they can get away with it. Stay one jump ahead by varying details in the assignment. Part of our job is to make it so difficult to cheat that it isn't worth the effort.

Field Trips

Another "real-world" application tool is the field trip, visiting a public service or commercial activity to expose students to the details of the operation, explained by those in charge. Properly planned and executed, these are excellent learning experiences. Like demonstration, they prove that the concepts you have been discussing in class actually work. Within the fire service, probably the most commonly visited activity is the local dis-patching center. Among commercial facilities, fire service classes routinely

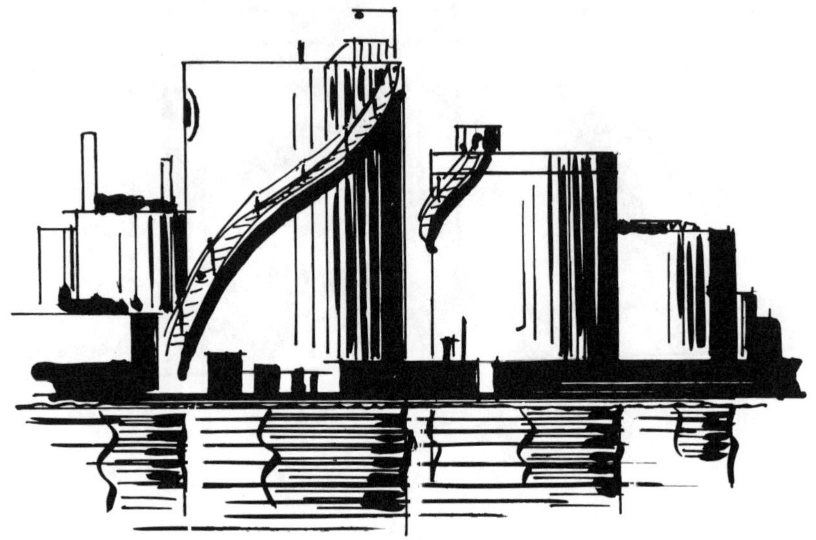

Fig. 7-4. Field trips enhance learning.

visit bulk oil and gasoline storage plants, liquified petroleum gas facilities, and similar target hazards.

It is important to lay some foundation with your students before the visit, giving them an idea of what to expect and what to observe (Fig. 7-4). You should also require some form of post-visit participation on their part. If you do not follow the field trip with a required report or, at least, some active class discussion on what they saw and learned, much of the value is lost.

Field Assignment Protocol

In arranging your facility visits and inspection assignments, use common sense and courtesy. Make your requests with plenty of advance notice and give the managers some flexibility in setting the time and date of the visit. Avoid peak work periods, shift changes, or any other times which might be inconvenient, just as you would in planning a routine fire department inspection. It is a good idea to confirm your arrangements in advance, in writing, and a better idea to write a thank-you note for the courtesy that has been extended. A complimentary letter to the superior of the person who served as your host or guide, commending the individual if he or she has been particularly helpful, is a nice gesture. Remember to send a copy and a personal note of thanks to the individual involved. Courtesy pays, particularly if you need an invitation to return to the same place with another class or group.

Role Playing

An application technique especially useful in areas of managerial instruction is role playing. The most common scenarios are handling disciplinary situations, resolving rules confrontations, dealing with problem workers, or presenting and gaining acceptance of some operational change before top management. There are other applications, as we will see; you can "role play" almost any form of interpersonal encounter.

As the name suggests, students are provided with a fictional situation or problem, and they are asked to "act out" the role of the leader, manager, or instructor, applying principles or concepts which were taught in class. Their "co-actors" may be other students in the class, other firefighters or officers borrowed for the occasion, or instructors. We are inclined to suggest the *class* instructor *not* become directly involved in the role-playing situation, unless the subject area or situation chosen demands some special technique or finesse. The reason, clearly, is that it is difficult to evaluate student performance while you are engaging in the intense give-and-take which effective role playing requires.

Some Difficulties

Developing and controlling a truly educational series of role-playing opportunities is not easy. To capture student imagination and interest, the scenarios or situations you present must be believable and obviously important to the subject material under study. At the same time the situations must be simple and clear, so the key elements may be quickly grasped without a great deal of lengthy introduction and explanation. Your students must be highly motivated and they must be willing to put themselves into emotionally volatile situations, understanding and accepting the fact that there is an element of hazard to the egos of those involved. Participants will often become totally frustrated, or hurt, or angry. As instructor and friend, you have a serious responsibility to ensure that the process is at all times *constructive*—never destructive. For role playing to have any merit as a learning experience, this high level of personal involvement is absolutely necessary. On the other hand, if students merely walk woodenly through an exercise or treat it as a joke, role playing is essentially worthless.

Fig. 7-5. Role playing at Dorking.

An Effective Use of Role Playing

One of the best applications of role playing we have seen was carried out in the late 1970s at Dorking, then the British Fire Service Staff College. In a part of the curriculum concerned with public relations and dealing with the media, fire officers were required to role play a television interview (Fig. 7-5). Each day, just before the lunch break, one class member was given a scenario of a particularly difficult public relations problem. These scenarios included such problems as a disaster with mass casualties and muddled lines of authority, for example, or a fire situation in which firefighters died under circumstances which appeared to be avoidable, or perhaps a fatal fire at a country club in which the victims, well-known sports and film personalities, died as a result of alcohol or drug intoxication. Each problem was designed to establish a situation which demanded a public statement from the senior fire official, and, at the same time, created a public relations "mine field" through which the officer had to thread his way with utmost care.

After lunch, the officer, in full uniform, was put into an appropriate indoor or outdoor setting and confronted with a video camera Tr and an aggressive and obnoxious television news reporter, complete with trench coat and slouch hat, microphone in hand, played by one of the Dorking staff. With classmates back in the classroom watching on monitors, and the videotape recorder rolling, the "reporter" would do his absolute best to break through the composure of the "chief officer," who would, in turn, do *his* best to make the necessary statements without "destroying his career" in the process—a situation filled with all kinds of stress and ego-bruising potential.

When the role-playing exercise was completed and the student "chief officer" was back in class, the instructor would run the videotape. In this moment of truth, the individual involved could see exactly how well or how poorly he had handled the situation, and the instructor and students would discuss the positive and negative aspects of the performance. It took a lot of time and effort to prepare and execute, but it was an absolutely superb learning experience for potential chief officers. It was, as we said role playing should be, believable, relevant, quickly understood, and more than a little stressful.

Simulation

Simulation, in current fire service terminology, usually refers narrowly to some audiovisual form of tactical or strategic fireground training. We will talk about that in detail in just a moment. The word, however, covers a much wider range of learning applications. Webster's *New Collegiate* defines "simulate" as: "to copy, or to represent, or to pretend." Role playing, for example, is simulation. A practice teaching assignment during an instructor training course is simulation.

The adjective "simulated" gives us an additional important insight. It is defined as: "made to look genuine." Any form of fire training simulation *must* be *made to look genuine.* The hallmark of our simulations must be realism—careful, thorough, painstaking realism. In the Dorking example just described, the appropriate indoor or outdoor setting, the mandated full uniform for the student, the dress of the TV reporter, the ever-present microphone, and the video recording all contribute to the realism of the exercise; all help to make it look and feel genuine.

In all forms of simulation, details are extremely important. Details create a situation and an environment which give each participating student the maximum opportunity to *believe.*

By definition, our simulations cannot be real; they can't be diamonds, but they should be the very best imitation gems we are capable of producing. If we settle for anything less, we are cheating our students.

Applications for Tactics and Strategy

Over the past two decades many tactical and strategic simulations have been developed. While the hardware and the mechanics have varied considerably, the goal has been much the same; to create a realistic fireground decision-making process with enough "built-in" stress to place students in a pressure situation. In most cases, the pressure is developed by forcing the student "officer" to make rapid decisions in full view of the peer group and instructional staff.

We know from experience that it works. We have seen seasoned fire officers break out in a cold sweat, mentally wrestling with a momentary data overload, while their colleagues wait in happy and sadistic anticipation for them to give the orders which may, or *may not*, control the situation. If you have any doubt that a classroom simulation can generate that much stress, we urge you to try it sometime, in front of *your* peers.

Hook-and-Loop Board Problems

Boards of the hook-and-loop type, visual aid display panels of various sizes and colors on which graphics are held in place with hook tape, have been used successfully in creating strategic exercises. The board serves as a pictorial map on which the problem evolves as a changing series of graphics are displayed, clearly showing the location, scope, and nature of a large-scale emergency situation, perhaps a major aircraft crash in a populated area, a devastating explosion, or a complex transportaiton fire. The visual changes on the hook-and-loop board are orchestrated to be consistent with a series of paper messages, typical of the type of information and questions which would flow into the incident command post (Fig. 7-6).

In a series of Maryland staff and command courses, the prepared paper messages, as many as ninety for a one-hour problem, were carefully designed to force the designated incident commander to utilize a full range of staff positions, each "role played" be a fellow student. The complexity and flow of data was intentionally designed to exceed the capability of any one single individual and was also carefully arranged to provide a relatively balanced level of activity for each of the staff officers. Message information was intentionally not provided to the command post in the time sequence or in the order the simulated events would have occurred. Messages were arranged as the information might be expected actually to trickle in to the command center; confused, out of order, and sometimes in error.

The problems described, along with preparation and critique, served as an application step in the closing three hours of a twenty-four-hour staff and command course. As should be the rule in all simulations, the problems were permitted to run their course with minimum instructor input. Discussion of the accuracy of the decisions made and the methods employed was conducted during the post-problem critique; *not during the problem*—an impor-

Fig. 7-6. Staff and command simulation with paper messages.

tant point to remember! A problem cannot be believed, cannot appear genuine, if the instructor is constantly inserting comments or questions into the thought processes of the students. Teach during the critique, not during the problem.

Table-Top Tactics

At a more elementary learning level, small cardboard scale models of structures have been employed as the basis for table-top exercises in tactics classes (Fig. 7-7). Students visit and pre-fire plan specific structures in the community and give an in-class report on the construction, layout, contents, and fire problem which each building presents. Each student constructs a model of the building assigned to him or her in a scale appropriate for use with toy fire apparatus.

In the simulation exercise, the model building is placed on a prepared map which shows streets and water supplies. Exposures may be shown by the use of models or they may be drawn in plan view on the street map. Based on knowledge gained from the student's report, the instructor places tufts of cotton, appropriately spray-painted in flame or smoke color, on the model to indicate the scope and location of a realistic fire problem. Scale

Fig. 7-7. "Table-top" models provide inexpensive simulation.

lengths of string are used to represent hose, with different colors or different thicknesses designating hose of various diameters.

Using limited-range "walkie-talkie" radios or simple wire and speaker links, students role play the dispatch center, the officer in charge, and the company officers commanding the responding units. Miniature apparatus is staged into the problem in some predetermined-scale-accelerated time frame, such as one minute of real time represents three minutes or five minutes expended of exercise time (Fig. 7-8). Upon "arrival," unit officers place apparatus and hose lines according to orders from the "officer in charge" or as dictated by local standard operating procedure. The instructor causes the simulated fire to grow or diminish according to the effectiveness with which student officers apply correct and appropriate tactics. At the close of the exercise, a critique again provides an opportunity for correction, guidance, and reinforcement of teaching points.

Fig. 7-8. "Scale time" can easily be achieved by removing the hour hand from a standard wall clock and covering the face with a home-made dial indicating any scale time you choose. *Courtesy Maryland Fire and Rescue Institute.*

Projected Fire Simulations

In the early 1960s, the U.S. Forest Service (U.S.F.S.) pioneered in the development of equipment and techniques for creating moving fire and smoke on a projected background scene (Fig. 7-9). In those years, the U.S.F.S. pulled supervisory personnel from all parts of the nation to direct fire suppression efforts during peak wildland fire periods. Most of these people were regularly assigned to other duties. Their involvement in major fire suppression command responsibility, therefore, was intermittent and relatively infrequent. The goal of the Forst Service simulation research was to find a highly realistic means to train this cadre of "part-time" fire officers and to maintain and hone their fire suppression skills during the lengthy periods between firefighting assignments.

When you stop and think about it, that situation is not unlike *our* need to train the fire suppression officers in volunteer fire departments and in smaller career departments where major working fires are also "intermittent and relatively infrequent." For this reason, we believe projected fire simula-

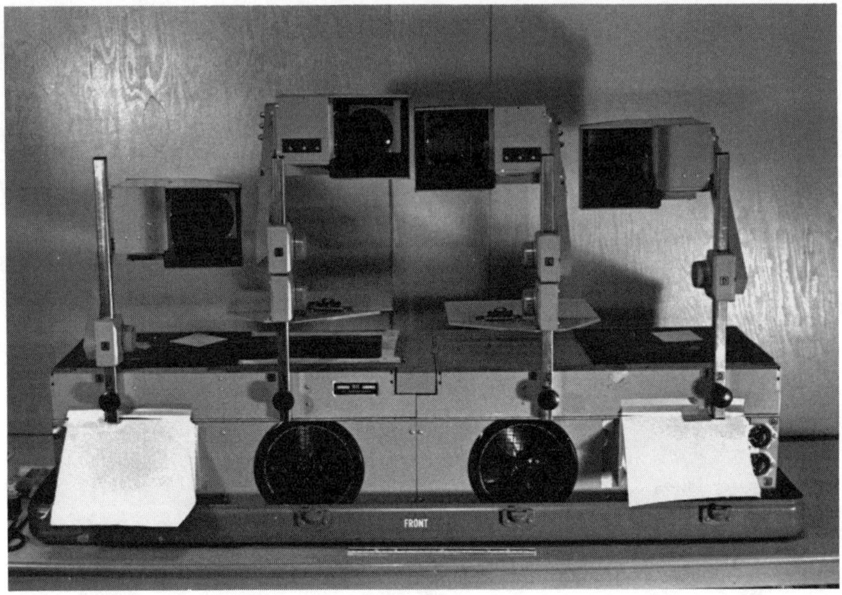

Fig. 7-9. U.S. Forest Service portable simulator, front view. *Technovate Simulator; courtesy Maryland Fire and Rescue Institute.*

tion is one of the most important tactical and strategic application tools we have available today. It gives our students an opportunity to apply command theory, to practice command skills, and to experience the stress of command without endangering life and property.

If the projected simulations are skillfully and realistically created, and if the fire scenarios accurately replicate "real-world" fire behavior, it is possible to achieve *total* student involvement. With good equipment, proper planning, painstaking execution, and the application of some judicious peer pressure, the intensity of this learning experience can come very close to actual fireground command.

Fundamentals of Application

In a learning situation using projected fire simulation, students are positioned in front of a dynamically changing screen on which is projected a fire problem they must control. Using a simulated communications network and maps for plotting fire progress and the disposition of apparatus and hose lines, students make and implement decisions, applying the principles of tactics and strategy which have been taught. Time may be actual or in some predetermined scale. The instructor in charge of the exercise, communicating privately with the simulator operator, causes the fire to extend or decrease according to the actions taken, or, in some cases, because an appropriate action has *not* been taken.

The Mechanics of Projected Fire Simulation

The staff of the Fire Service Extension, University of Iowa, were the first to construct a "home-made" fire simulator, based on the Forest Service research, and they were the first to apply the technique to structural fire suppression. Following in their footsteps, many fire training agencies have built simulation devices over the past twenty years. These locally assembled simulators are most commonly built around a pair of standard overhead projectors, one for creating fire and one for creating smoke (Fig. 7-10). Each is modified by the addition of some form of driven wheel, rotating in the beam of light, to create the illusion of motion.

The stage of the fire machine is covered with transparent red-orange plastic and the stage of the smoke machine is left clear or, in some cases, covered with transparent mottled gray. The stages of both machines are then blanked out with a sand tray or a glass plate painted flat black. Fire and smoke are produced at a specific spot on the screen by clearing a hole in the sand or by scratching a hole on the painted surface of the glass plate. The projection head is raised or lowered to establish an appropriately "fuzzy" image on the screen. Both fire and smoke are projected out of focus; how far out is determined by the background scene and the effect you are trying to create. Experimentation is the only teacher.

The background scene is projected with a 35-millimeter slide projector or from another overhead projector using an eight-inch by ten-inch transparency.

This form of projected simulation is widely used by fire training agencies in all parts of the country. Unfortunately, much of it is poorly done. While it might be argued that any simulation is better than none, we disagree. This technique is far too valuable to waste in any level of effort inferior to the best we can make.

"Made to Look Genuine"

Remember those words? "Simulated" was defined as "made to look genuine." By definition, then, merely squirting a blob of pink light on to a

  Fire Instructor's Training Guide

Fig. 7-10. Projected fire simulation. *Technovate Simulator; courtesy Maryland Fire and Rescue Institute.*

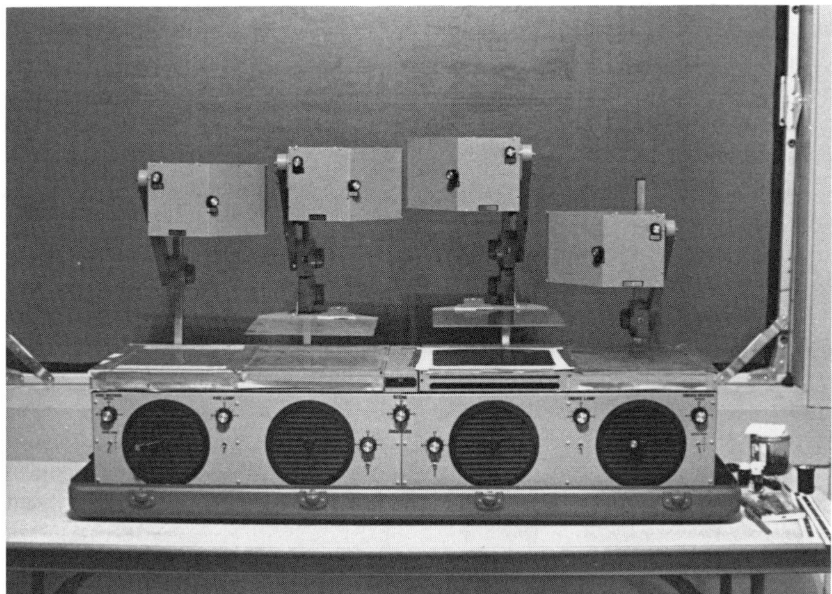

Fig. 7-11. U.S. Fire Service portable simulator, back view. *Technovate Simulator; courtesy Maryland Fire and Rescue Institute.*

projected picture of a building is *not* simulation. Don't settle for that level of performance. It is possible to make fire and smoke appear genuine with home-made equipment, but not without some expense and effort. Let's consider what is required.

In the late 1960s, in about the third or fourth generation of their simulation hardware, the U.S.F.S. contracted out the construction of a series of portable simulators for their own use and for purchase by other interested fire training agencies. Two of these machines are still in active use within the Maryland Fire and Rescue Institute (Fig. 7-11). The specifications for these units produced some excellent features, features worth noting and duplicating in any home-built simulator or specifying on a purchase order for a commercially built unit.

Essential Variables

The light levels for both smoke and fire projection are variable. The ability to vary the intensity of the light which produces smoke on the screen enables the operator to adjust the brightness of the smoke, according to the density of the projected background scene. Flame, on the other hand, is usually projected at or near the brightest level available. The speed and the direction of the wheels used to create motion in both smoke and fire can be changed (Fig. 7-12). These variations in intensity, speed, and direction are absolutely essential in creating realistic fire and smoke. Another key to creating a realistic scene is to use *irregular* openwork on the faces of the motion wheels to provide the irregular fluctuation of actual flame and smoke.

The amount of motion *we see* in an actual fire is a product of the amount of heat being generated, and it is also affected by the distance between us and the flames. The hotter the fire and the closer the viewing point, the more motion we will perceive. In addition, flame and smoke do not necessarily move at the same rate; fire motion is generally more rapid, while the speed with which smoke rises is dependent on the amount of heat being generated.

Variations in direction and speed are achieved by adjusting the placement of the motion wheel in the light beam and by using variable-speed reversible motors to drive the wheel. This flexibility permits the simulator operator to create smoke which appears to be moving rapidly or slowly, rising or drifting at a variety of angles to the right or left, according to the desired scenario.

On the fire projector, this same capability to change speed and direction permits creation of realistic flame, rolling out of windows, doors, or burned-through areas, turning upward, rising on the thermal column like actual fire. Thermal currents follow the rules of fluid dynamics, rising and flowing over and around surfaces just as water flows down. When you next see a good slide or photograph of a structural fire with flame pouring out and then turning upward, particularly if it is moving around some obstruction or across a surface such as a porch ceiling and then turning up again, turn the picture upside down, and you will see exactly what we mean. The fire will appear to be flowing just as a stream of water flows over and around rocks and cascades down a waterfall. This appearance is the type of "genuine" movement and flow we want to simulate.

Sand Trays versus Painted Glass Plates

We feel it is practically impossible to achieve this degree of reality by using sand trays to block out the fire and the smoke screens. It is extremely difficult accurately to control the size and shape of the fire image on the projection screen by pushing aside sand, and there is always the possibility of an accidental movement of the tray which will change the size and shape of

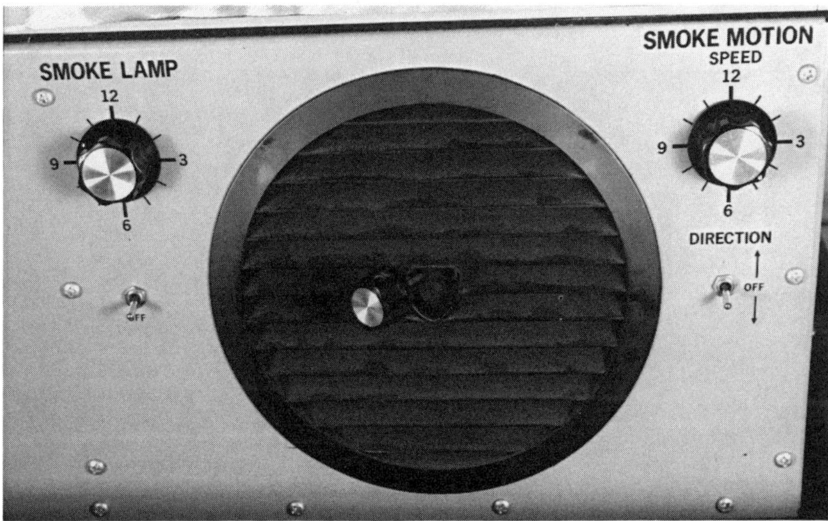

Fig. 7-12. Lamp controls alter intensity of light. Motor controls alter speed and direction of motion. *Technovate Simulator; courtesy Maryland Fire and Rescue Institute.*

 Fire Instructor's Training Guide

Fig. 7-13. With glass plates, precise placement is possible. *Courtesy Maryland Fire and Rescue Institute.*

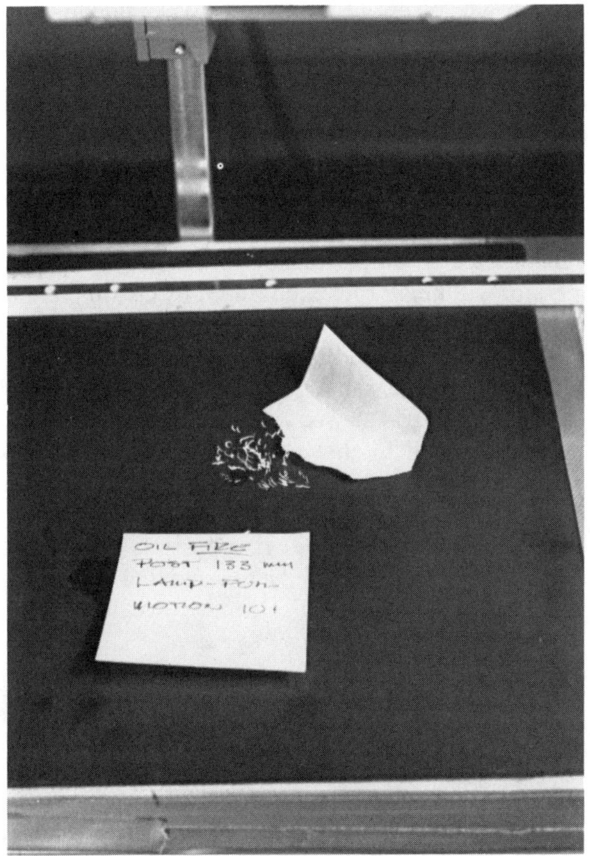

Fig. 7-14. Torn pieces of paper, moved realistically, permit the operator to expand or contract the fire. *Courtesy Maryland Fire and Rescue Institute.*

the projected flame or cause it to "go out" completely. Glass plates, sprayed with flat black paint, can be easily and accurately scratched with a pointed scribe to place fire and smoke exactly where you want them (Fig. 7-13). Working with care, and making extremely small holes or fine cuts in the paint, it is possible to create the illusion of fire *inside* a window on the projected structure, or behind a railing, or just breaking through at the eaves. The level of control this method provides is excellent. You can place the fire exactly where *you* want it.

Working with black-painted glass plates also gives the simulator operator the capability to make up problems in advance. We use a simple form to note variations in the set-up; we record the height of the projection heads above the stage, to ensure that we regain the proper "out of focus" appearance in the fire and smoke, and we record the lamp and motion settings. Each pair of plates, smoke and fire, are labeled and keyed to the background slide. With prepared plates and a sheet of notes on the simulator settings for each scene, it takes only a few minutes to recreate and "fine tune" any particular problem when you are ready to use it in class.

We also plan in advance any anticipated extension in the fire or increase in smoke. This can be accomplished by identifying the spot on the plate where the operator will begin the extension or by creating the extended fire and covering it with a scrap of paper when the scene is initially displayed (Fig. 7-14). Fire is "extinguished" by this same method: blocking out the flame with bits of torn paper, the ragged edges used to reduce the size and scope of the fire gradually and slowly, without projecting a harsh straight line across the flame front—all in the interest of "making it look genuine." Permanent corrections are made with a small brush and flat black paint. In this same way, the plate can be filled in for reuse, or, if the unwanted scene is extensive, the plate can be resprayed. It is possible to reuse plates quite a few times before they must be cleaned and repainted.

The Simulator Operator

Turning once again to Webster, we find that a "simulator" is defined as "a device that enables the operator to reproduce or represent, under test conditions, phenomena likely to occur in actual performance." What an appropriate definition for our purposes! It also brings into focus the role and importance of the operator. From our experience, a fire simulator operator does *not* need to be an artist. The operator *does* need a thorough knowledge of fire behavior and an eye for the appearance of flame and smoke under varying fire conditions. A knowledge of building construction and the effects of building construction on fire progress are also essential (Fig. 7-15). The operator, therefore, should be a person with fireground experience, a good eye for detail, and enough artistic ability to translate this knowledge into a realistic scene. The simulator operator must coordinate every action with the instructor or "umpire/director" who is conducting the problem and cause the fire to grow or diminish according to that person's instructions.

Rear Projection versus Front Projection

We feel rear projection is essential to creating a realistic illusion in projected simulation. We recommend the relatively inexpensive flexible

Fig. 7-15. The simulator operator must know how fire moves. *Courtesy Maryland Fire and Rescue Institute.*

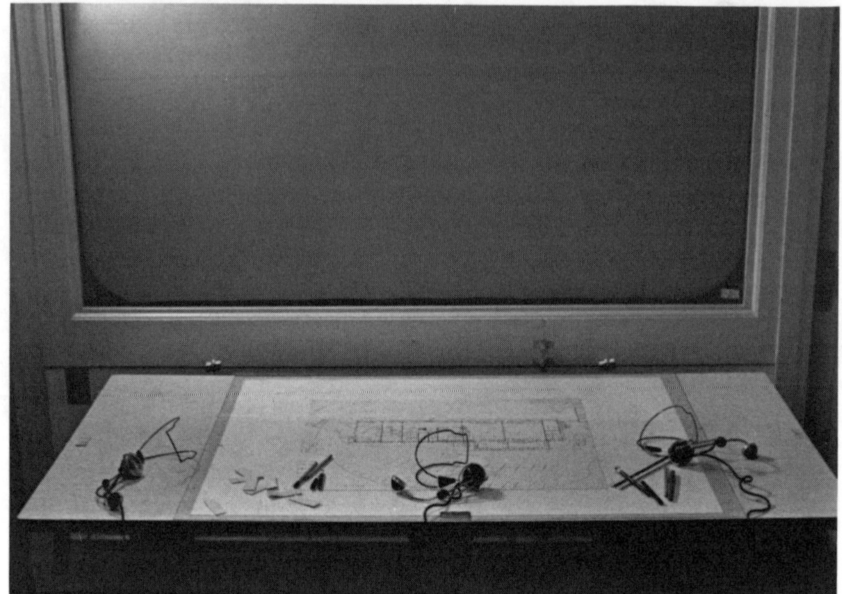

Fig. 7-16. Rear projection screen and student "game board" with hard-wired communications simulation. *Courtesy Maryland Fire and Rescue Institute.*

plastic screens which snap on to a metal framework. The best possible arrangement is to have the rear projection screen built into a wall which separates the simulation room from the room in which students are working the problem. The simulation room can also be used to separate the mock dispatch center from the area in which students are working. This arrangement makes some form of communication net absolutely essential, including a private communication link between the simulator operator and the person supervising the problem. An alternative is to use a portable rear projection screen with wing panels or screens which separate the two areas.

In either form of rear projection, the space immediately in front of the screen can be used for a table to hold maps, status boards, or any other form of display that you provide for students to track the implementation of their solutions (Fig. 7-16). Student attention should be focused, without distraction, on the development of the fire and on their "game boards." Standing behind the student "officers," the problem supervisor has a clear view of both the screen and the solutions being generated. From this position, it is possible to assess the effectiveness of student tactics in relationship to the problem on the screen and to communicate quietly to the simulator operator any changes you want to make in the projected fire and smoke.

With rear projection, room lighting levels on both sides of the screen are critical. We suggest that room lights be turned off. This darkness forces us to provide some form of low-level lighting, shielded from the screen, over the table space the students are using, and for the dispatch operator, if that person is in one of the darkened areas. Also, ensure that no stray light beams from the projectors strike the screen. A piece or two of strategically placed poster board will usually solve that problem.

Front projection puts the simulator equipment right in front of the screen, in the same space where students are working the problem. This method can be used, and it has been used in many classes, but it is certainly not a desirable arrangement. Imagine trying to watch your favorite movie, in a theater, with the popcorn stand open for business between you and the screen! You might see the picture and hear the words, but it would be just about impossible to concentrate your attention on the illusion of reality which the film was designed to create. Exactly the same situation exists when the presence of the simulation equipment and the simulator operator, right there in the middle of everything, is a constant reminder that this is only a game. We just don't think it is a good idea. Like using poorly developed equipment or carelessly created, unrealistic fire and smoke, it is another misuse of an extremely valuable application technique.

Adding Light—Subtracting Light

With the projection techniques just described, we can only add light to the screen. This limitation has two disadvantages.

First, the background scene must be fairly dark for fire and smoke to show up effectively. The simplest way this can be achieved is by taking a series of three or four slides of each scene you want to use, beginning with a normal exposure and then either increasing the shutter speed or stopping down the lens opening, one additional increment as each picture is taken. This technique will provide a set of slides of varying density from which the simulator operator can pick the one which works best. As an alternative, it may be possible to control the slide projection lamp with a rheostat which bypasses the cooling fan and permits the fan to run at full speed while the current to the bulb alone is reduced to darken the scene.

The second disadvantage is that only light-colored smoke can be projected, since we can only add, not subtract, light. This limitation is not a great problem in creating night scenes, but it is definitely unrealistic in setting up a daytime scenario or a fire involving petroleum products. The

Fig. 7-17. Simulation using eight-inch by ten-inch transparency. *Technovate Simulator; courtesy Maryland Fire and Rescue Institute.*

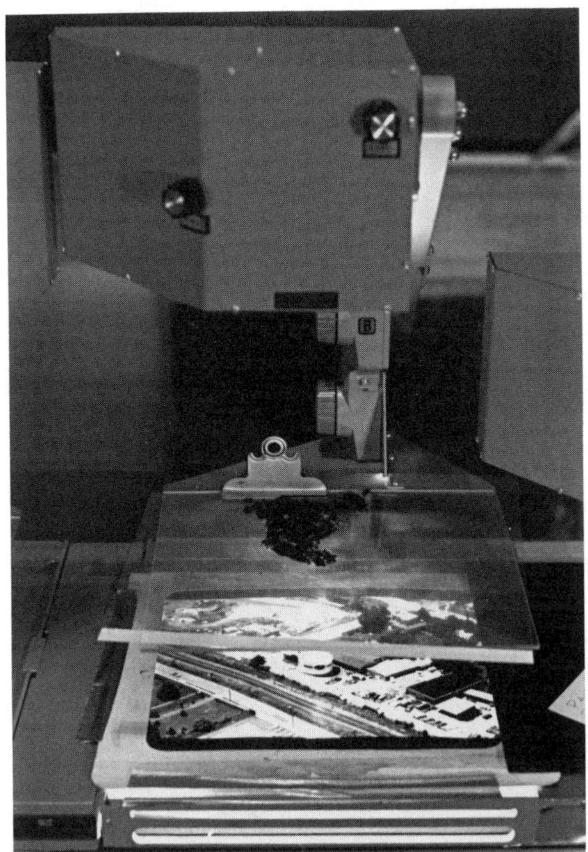

Fig. 7-18. Intermediate "out-of-focus" stage permits "painting in" smoke on a clear plastic sheet. *Technovate Simulator; courtesy Maryland Fire and Rescue Institute.*

only way we can subtract light, or project "dark," is to interrupt the beam of light projecting the background scene and throw a shadow on the screen. On the U.S.F.S. portable simulators, this action was accomplished by using an eight-inch by ten-inch transparency to create the background scene and by mounting an intermediate stage, a sheet of clear plastic, between the main stage of the projector and the projection head (Fig. 7-17). The intermediate stage is overlayed with an acetate sheet on which the operator can "paint out" unwanted light, creating dark smoke or "char" on the projected image (Fig. 7-18). This method is an excellent arrangement, and it is relatively easy to use. The disadvantage, and it is a significant disadvantage for most of us,

Fig. 7-19. Adding "smoke" to 35-millimeter projection. *Courtesy Maryland Fire and Rescue Institute.*

is in the cost of the large transparencies, which can run as much as $30.00 each.

The same effect can be accomplished with a 35-millimeter projector, although with considerably more difficulty, by building a wire framework, arranging it around the projected light beam, and clipping bits of paper or cardboard on the wire frame, creating shadows on the screen where desired (Fig. 7-19).

The "smoke" shadow can be filled with extremely realistic motion by using a low level of light from the smoke projection lamp and appropriate swirls scratched on the smoke plate. An excellent illusion of a flammable liquid fire can be created by having flickers of flame roll up into the smoke cloud, just as they do in reality. *Go for it!*

We have spent a lot of time talking about projected simulation because it is, in our opinion, one of the most important fire service teaching-learning tools. There is a wide range of commercially available simulators with

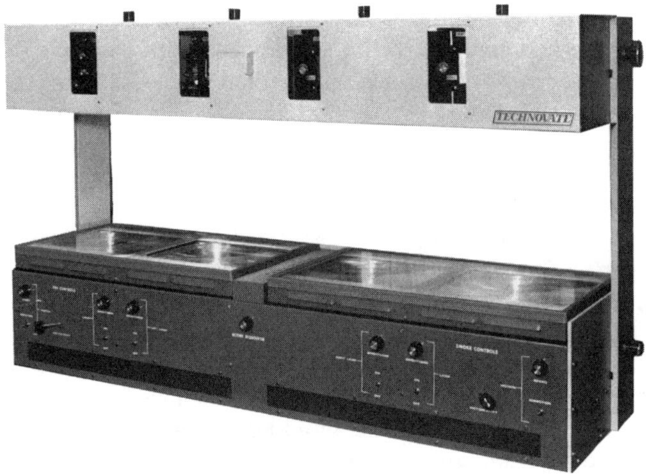

Fig. 7-20. Situation simulator, current technology. *Courtesy Technovate, Inc.*

varying capability (Fig. 7-20). We also believe, knowing as we do the number of superb "shade tree mechanics" and part-time inventors in the American fire service, that the necessary equipment could be created in even the smallest training agencies. If you are responsible for training fireground commanders and you are not using effective, realistic, projected simulation, you should be. Do it!

And Then There Are Computers

As this book is being written, computer-driven fire tactics and strategy application games are just beginning to enter the market. We have not, in all honesty, seen any which we can wholeheartedly approve. The ones we *have* seen are rather narrow in scope, and they call on the student to provide only a limited range of decisions. The graphics are relatively crude and require a lot of student imagination—much more imagination, certainly, than is required in projected simulation.

There is no question that the technology is there; all that remains is for someone to develop realistic and comprehensive programs. How soon that will occur is mostly a matter of economics. Compared to other markets, the fire service is so small that it does not encourage the heavy investment which any form of computerized instruction requires. We may have to wait a considerable length of time, or there may be a program of excellence right around the corner. Keep a keen eye on developments in computer technology, keep an open mind, and be ready to jump in and experiment when a really effective computer-driven application tool appears. Or, if you have an ample knowledge of programming and if you are a keen student of tactics and strategy, get busy.

Tomorrow, Who Knows?

None of us can really imagine what fantastic and exciting developments will happen in fire service education and training over the next several decades. When the National Fire Academy was still at 2400 M Street, in Washington, we attended a meeting exploring simulation with one of the leading aerospace manufacturing firms. They were talking about creating a fire training scene using computer-driven holography, projecting a three-dimensional, full-color image of a building through which a simulated fire would extend according to the building construction and layout data fed into the computer program. They assured us that the technology was ready and waiting, that their firm was ready and waiting; all that was needed was money. The price? Mega-millions. Nobody even blinked. In those days the sky was the limit, and nothing was too good for the fire service of our nation.

Well, that bubble has long since burst, along with a lot of other dreams we had, but the technology is still out there, somewhere. Equally exciting ideas will someday be a routine part of your training program, or perhaps the one that follows yours. It will certainly be great fun to work with, but the most important and irreplaceable element will still be *you,* the caring human being at the front of the classroom.

Application Steps on the Drill Ground

We routinely "simulate" real-world conditions, or close to real-world conditions, on the drill ground. We require students to apply learned skills on the driver training range, at flammable liquid or flammable gas props, in smoke houses, purpose-built structural fire buildings, or in real buildings burned for demolition. These techniques are so well known and so widely used that we will not waste your time in reviewing them here. Our prime concern for this part of your work, again, is safety.

In our search for "realism," it is very easy to get carried away. We repeat our earlier caution that you *do not innovate* during class. Any additions or changes to hazardous training evolutions should be carefully planned, and the procedures must be tested by instructors or other experienced personnel

Fig. 7-21. Training for working in total darkness.

before students are permitted to try them. This practice should be an unbreakable rule in any training program, particularly *yours!*

Tips and Ideas

We *will* share a few concepts we have seen and liked, application tips and ideas which add realism or make your job easier. "Bowl covers" for blanking out the facepiece of a breathing apparatus, for example—we have seen these made of translucent "frosted" plastic, or made of black cloth, with an elastic band sewn into a circular hem which fits snugly around the facepiece rim (Fig. 7-21). They are easy to put on and take off, and they work a lot more effectively than the masking tape which most of us use.

Blanking out the facepiece to permit a novice firefighter to learn to move and work by feel alone is a common application technique. One easily arranged indoor or outdoor drill is to lay out a path with several sections of charged or uncharged hose line, crossing the hose over or under itself a time or two and passing it over or under available obstacles like chairs or tables or hose racks. The student, with mask on and facepiece blanked out, must follow the line from the nozzle to "clear air" by feel, keeping one hand on

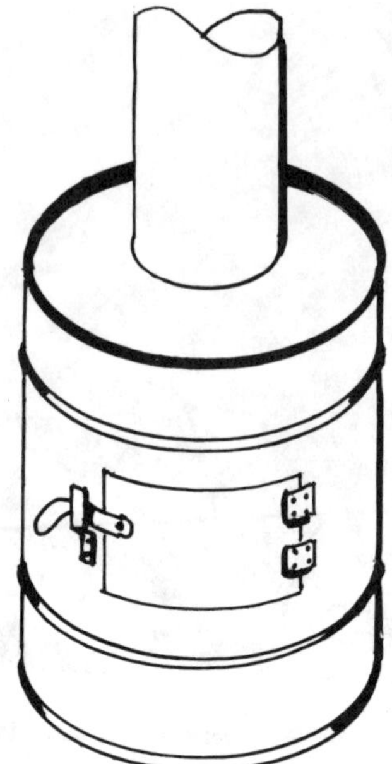

Fig. 7-22. Just the opposite of ventilation!

the hose at all times and identifying, at the instructor's request, the items encountered.

Moving up to the next level of breathing apparatus training, usually in some form of maze or "smoke house," consider installing mock-up utility props like electrical pull switches and gas meters. These provide some realism, they give the student something to search for, and they can be easily checked by an instructor to see if the student actually did find them and turn them "off" as directed. They also provide an opportunity for the student to achieve one of those little "early successes" we talked about back in Chapter 3.

One useful "smoke house" technique is to burn wet straw or excelsior in a simple stove, made from a small drum, set up on the *outside* of the building with the smokepipe leading *inside*—just the opposite from the way in which sensible folks use stoves (Fig. 7-22). This device gets the hot stove out of the structure, and it makes tending the fire and adding fuel a much easier and safer task than when the smoke-producing arrangement is inside the building.

When a purpose-built structural fire training building is available, consider running some combined firefighting and rescue simulations. The search team, coordinating with the hose line team, must locate and remove rescue dummies or firefighters who are "victims." We do *not* recommend this practice as a drill when you are burning an actual building. There is always a chance the fire will get away, and it would be all too easy to lose a firefighter hidden in some corner.

Disaster Drills

Disaster drills are easily the largest simulations in which we participate. In all honesty, we have rather a poor opinion of disaster drills—most of them, anyhow. Too many of them live up to the name and are actually disasters in themselves. Because of the large numbers of personnel and the large quantity of apparatus involved, these exercises require a tremendous amount of planning, preparation, logistical expertise, and cooperation. Failure in any one of these areas causes the whole house of cards to come down. That's when you discover that people who feel they have wasted a Saturday or a Sunday can get downright hostile.

A major disaster drill, properly implemented as a teaching-learning experience, is partly application and partly evaluation. To have any merit, and to have the best chance of success, the drill should come at the end of an organized training program, centered on the activities and skills which a major disaster requires. The participants should be applying learned principles and the supervising instructors should be evaluating their ability to do so. This seldom happens.

More often a drill is put together to satisfy some regulation or statute which mandates that drills be held periodically. Because of our expertise and experience, state and local fire instructors are pulled in to provide assistance, advice, and counsel. When things go wrong and the participants are looking for somebody to throw rocks at, there we stand, front and center—not a comfortable place to be.

Properly used, properly planned, and properly executed, a disaster drill can be an excellent educational tool. We can only urge you *never* to underestimate the difficulty of organizing and implementing a training exercise of this size and scope. It ain't easy.

Application and Training Aids: How Do They Fit Together?

We've been talking in this chapter about participatory experiences, experiences and activities in which our students can apply the concepts and skills we are conveying to them. Some of these techniques, like projected simulation, clearly rely heavily on audiovisual aids. Some, like role playing, require a high degree of emotional involvement.

Let's move on to the next chapter and the whole scope of audiovisual materials and equipment available to us, and we'll begin by relating the material or the technique to the level of emotional involvement it requires or causes.

Come down the hall to the audiovisual layout room. That's a place where we can really put our imaginations to work and make use of many of our personal talents. If you enjoy being creative, if you like to work both with your head and your hands, it can be a lot of fun.

8

Training Aids

Over the years, we have seen several excellent descriptions of the range of student involvement that can be obtained through different kinds of learning experiences. As we continue to talk about training aid types and concepts, a subject which we wandered into during the preceding chapter with the discussion on simulator hardware, here is some good foundation information to file away in your mental computer.

Learning experiences range from completely passive to totally active. At the passive end of the scale, the student merely receives information, seeing or hearing only, taking no overtly active part in the process. At the most active end of the scale, the student is fully and productively involved, participating mentally, physically, and emotionally.

Beginning at the passive end of the range, we come first to those training aids and teaching techniques which require the student to use only a *single sense*. These include: *printed words*, like books, training manuals, articles, and reports; *spoken words*, like lectures and audiotapes; and *pictorial displays*, like drawings, diagrams, charts, and graphs. Above single-sense learning experiences, somewhere in the lower middle section of the passive/ active scale, are *multi-sense methods and materials*, beginning with illustrated lecture, demonstration, basic individual skills instruction, and such *audiovisuals* as motion pictures, videotapes, and sound-slide programs. Moving toward more active involvement are the low-key *group activities* like discussion, instruction in team skills, "buzz groups," field trips, projects, and project reports. Finally, a high degree of activity and emotional involvement are achieved in those experiences which require *intense personal participation* such as the role-playing and incident command simulations described in the previous chapter.

If we accept the validity of the "law of intensity," we must assume that the more active and the more emotionally involved we can keep our students, the more they will retain. We don't doubt that this theory is true, but you usually do not want to, and probably cannot, maintain so high a level of emotional intensity in every minute of every class. *All* of the teaching methods and materials are useful and each of them has a place, from those which require very little emotional commitment to those which require the student to take a significant personal emotional risk.

Since we have already talked at length about group activities and intense individual participation in Chapter 7, let's begin by considering the single-sense and multi-sense training aids available to us.

Training Aid Types

Training aids can be categorized in a number of ways. One simple method is by type; aids may be graphics, projections, or they may be three-dimensional.

Graphics

A graphic is any expression of an idea which is written or drawn on a two-dimensional surface (Fig. 8-1). Then, by definition, graphics include all written materials. We will take the liberty of not covering development of the written word. Creative writing is a field which is far too wide for us to examine in this text; it is also an area of study in which local training is available in community colleges or other adult educational programs. If your writing needs improvement, look around for a creative writing course and sign up.

Preparation of Graphics

A couple of pieces of poster board and a handful of felt-tip pens are enough to put you in business for constructing some useful graphics. Even

Fig. 8-1. Graphics.

with all the "store-bought" training aids we have today, there is often a need
to create some visual device to make a particular point or to fill a gap in the
prepared material. In teaching instructor training over the past two decades,
we have required our students to make their own graphics for use in practice
teaching the illustrated lecture method. We have never been disappointed
with the results, and we have often been amazed at the depth of talent and
ingenuity our instructor candidates have displayed.

In making your own graphics there are three key words to keep in mind:
big, *bold*, and *simple*—big, so the most distant student can see the smallest
detail; bold, so your words and drawings stand out vividly; simple, so your
message is clear and unmistakable (Fig. 8-2). Follow, too, the guidelines we
laid down when we discussed the illustrated lecture. Graphics should be
attractive, cleanly and artistically arranged. Use color for emphasis and
effect, but have a purpose. Making every letter on a graphic a different color
is a common mistake of student instructors. It actually does nothing but
make the words confusing and difficult to read. Color is highly effective, but
only if it enhances your message and does not distract from it.

Remember to limit each panel to a single concept (Fig. 8-3). If it is at all
possible, each graphic you display should draw the attention of your students
to the one single point you want to make at that time. What happens when
you display a graphic or a projection which contains a multiple series of
points, perhaps six or eight or ten? While you are explaining and expounding
on point number one, where is the attention of the class? Obviously, somewhere ahead of you as they read down the entire display. Finally, remember

Fig. 8-2. Key words for creating graphics.

accuracy. Make sure all facts and numbers are accurate and all words are spelled correctly. If you don't find your "misteaks" ahead of time, we guarantee your students will find them for you.

Construction Materials

Construction materials include just about anything you want to use: poster board, pencils, felt-tip pens, lettering guides, tranfer ("rub-on") letters, rulers, construction paper, poster paint, brushes, rubber cement, staples, brass split-leg fasteners, imagination, ingenuity, and patience (Fig. 8-4).

If you have trouble with lettering, there are a number of options. Lettering guides, cut-out templates of the alphabet and numbers, are available in a variety of sizes in most stationery or art supply stores. With these you trace the outline of each letter in pencil, and then fill it in with a felt-tip pen or poster paint. Don't leave the little breaks which are created by the template. Your work will look more professional without them. Rub-on letters are also available in a wide range of sizes and print styles. Each letter or number is transferred from the backing sheet on to your graphic by rubbing or "burnishing." Carefully applied, these look absolutely first-rate. With lettering all hand-done, correct spacing is established by keeping a uniform *area* between the letters, not by making the *distance* between letters uniform (Fig. 8-5). Different combinations of letters will have different distances between them if the area is kept constant.

If your ability to draw is limited, use an opaque projector or an overhead projector to project the art work you want on to blank poster board and trace the necessary outlines with pencil. You can then rework the "drawing" with felt-tip pens or paint to provide visibility and color. Using this technique, we

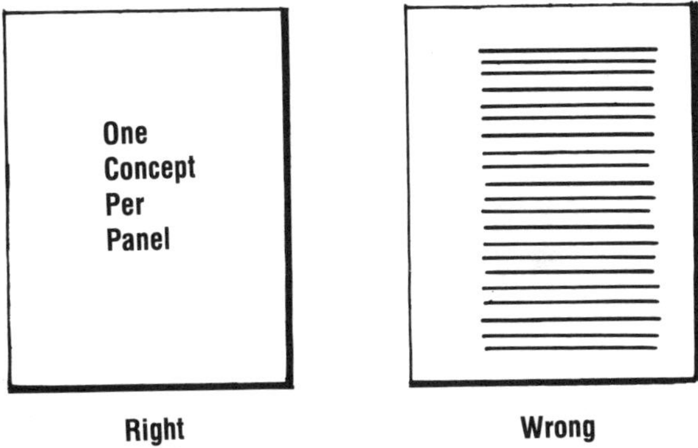

Fig. 8-3. Single panel—single concept.

have made "cutaway" poster board graphics of both wet and dry sprinkler valves, complete with movable parts mounted on brass fasteners. Working with a "hook-and-loop" type of board, each separate part of the sprinkler valve was displayed and discussed before the next piece was added (Fig. 8-6). By the time the whole complex assembly was complete, students knew the name of each part, how it functioned, and why it was there. Following

Fig. 8-4. If it works—use it.

SPACE

Fig. 8-5. Spacing between letters is based on area.

the rules we have just outlined, the drawings were big, bold, simple, attractive, accurate, and visible, and the display was put up one single piece at a time. The whole business cost only a few dollars and a couple of evenings of work to create. This technique could be applied to any device or equipment you must teach: pumps, pump primers, exhausters, accelerators, breathing apparatus demand valves, extinguishers, or anything mechanical.

Strip Signs or "Slap Sticks"

Strip signs are graphics in which each word or phrase is printed on a separate, narrow strip of poster board or similar material. This method permits the instructor to follow the "single panel/single concept" rule and still develop a lengthy list of items in the same display. The name "slap sticks" comes from the fact that the individual signs can be "slapped" against the board. (We suggest you put up the strips with a little more care to maintain a neater alignment and spacing.) Use a small piece of the "hook" tape on each end of each strip (Fig. 8-7). If you put only one piece in the center, it is almost impossible to get the strips to hang straight. It's also a good idea to number the strip sequence, or any sequence of graphic aids, on the back so that you can check to be sure they are laid out in the correct order. This technique also allows you keep them face down until they are displayed, if you wish to do so.

The strip sequence becomes part of your teaching plan, leaving you free to expand on each word or concept as it comes up. At the end of the list, if you take the strips down in the same order they were put up, you can summarize as you go, providing extra reinforcement of the points covered.

Poster board strip signs or any other poster board graphics will quickly get dirty unless they are treated, in some way, to preserve them. You can use them longer if you cover the face of each item with clear adhesive-backed plastic or spray them with artists' fixative. With spray fixatives, follow the instructions provided by the manufacturer and work in a well-ventilated area.

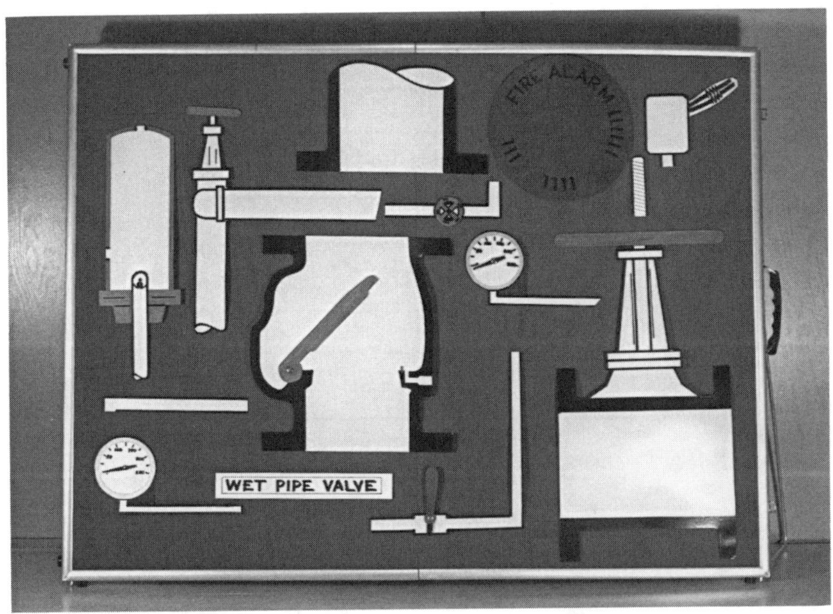

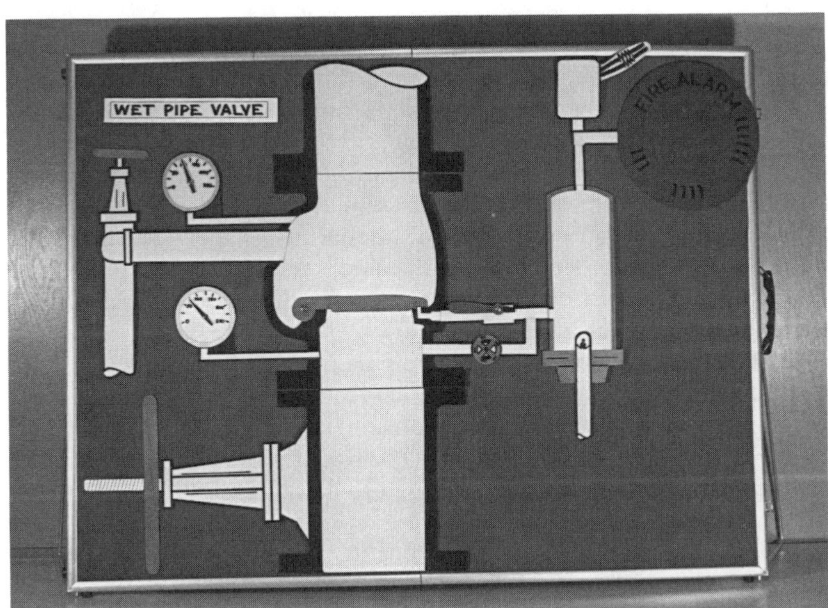

Fig. 8-6. Poster board hook-and-loop display.

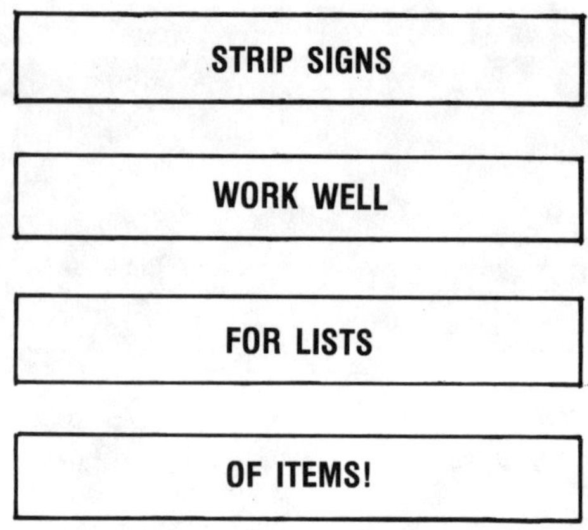

Fig. 8-7. "Slap sticks."

Display Boards

We have mentioned boards of the "hook-and-loop" type several times because we feel they are useful teaching tools. The boards come in a variety of sizes and colors, and some will fold for easy transport. You can also buy the "loop" material by the yard and make your own.

The "hook" tape comes in strips or rolls about an inch wide. You can buy it with an adhesive backing covered with a peel-off strip, or you can buy it plain and glue it on to whatever you want to hang on the board. The adhesive-backed tape works very well with poster board and other light-weight graphics. Cut the tape down the center and then cut across to make patches roughly a half-inch square. You don't really need that much tape to hold up even a large piece of poster board, but any smaller size is difficult to work with. The plain tape, used with contact cement or some other strong adhesive, works better if you want to hang something heavier, such as small objects or parts of a piece of equipment you are discussing. One square inch of the hook tape will hold approximately six pounds of weight on the board, so you don't need much. We once had the great pleasure of seeing an experienced and rather self-important instructor put up a good-sized graphic whose entire back side he had carefully covered with hook tape. It worked fine—until he tried to take it down. He couldn't budge it!

Most of the suggestions we have discussed concerning hook-and-loop boards will also work with magnetic boards. You can buy adhesive-backed magnetic strips and apply them just as described doing for the hook tape. The main difference is that the magnets usually won't hold as much weight.

If you overload them, they have a tendency to slide down, and that can be a little disconcerting when you are in the middle of some important point.

If you are working with a portable display board, make sure it is completely stable and solidly set up. Check before class and see that the board is visible from all parts of the seating area. Finally, check the spacing of your graphic layout. If you haven't done so already, make a practice run before class to be sure that all the pieces are in order and to be sure that the completed display will fit on the board. It is often a good idea to mark your starting point in some inconspicuous way to ensure you get the spacing you want when everything is up.

Flip Charts

Flip charts are excellent for keeping conference or discussion notes because they provide a complete record of what occurred, where notes made on a chalkboard would be lost (Fig. 8-8). These charts can be used for review, or summary, or serve as a base for transcribing the thoughts and ideas which were developed.

As teaching tools we felt flip charts were, at best, a poor substitute for a chalkboard, primarily because the size of the writing surface is so limited. Then we saw one instructor, a career Air Force sergeant, who could literally work magic with one, using extra-wide felt-tip markers, working over a white wax crayon base to create "invisible" words which appeared when the streaks of marker color were laid down, and writing white words on a marker color base using a cotton swab dipped in bleach. Instead of seeing limitations, he saw a teaching tool of great versatility and created enough variations that he filled the best part of an hour at an audiovisual seminar with the audience hanging on every word. There's a message there for all of us.

Chalkboards

Chalkboards in black, brown, or green still fill the front walls of many classrooms. We like having a big writing surface available to highlight a key word or to make a quick sketch to clarify a point; we don't think it matters much whether it is one of the traditional blackboards or one of the newer white writing surfaces we mentioned in Chapter 2. Both are effective teaching tools, and the choice is pretty much a matter of personal preference.

We keep a box of soft white chalk, a box of soft colored chalk in the deep, rich colors, and a well-used eraser in our teaching kit when we are working "on the road." Soft chalk because it makes a bolder mark than "dustless" chalk, the rich colors because we don't like the pastel tones of

Fig. 8-8. Old-fashioned but still going strong.

most "dustless" colored chalk, and a well-worn eraser because new ones don't work. While we are talking about using chalk, we should probably mention the squeaking or chattering which sometimes occurs. If either or both are a problem, change the angle at which you are holding the chalk or vary the pressure against the board. Soft chalk is less likely to squeak, but it sure will mess up a pair of dark blue uniform pants in a hurry.

As with all training aids, consider the viewpoint of your students. You must often limit your use of either of the boards, black or white, to the upper half or two-thirds of the surface to ensure that people in the back rows can see. Print, write, or draw legibly and make your words or drawings big. As they say in the contest rules, "Neatness counts"—take a few extra seconds and give your work the best appearance you can. Boards should usually be used for showing fairly brief material (Fig. 8-9). If you must use one of the writing boards for anything lengthy or complicated, put it on the board before class, to save class time. Most instructor training courses teach, "Talk to the class, not the board." We think it is a better idea to keep talking while you are working at the board, explaining what you are doing, defining the word you are writing, keeping up the flow of your verbal instruction rather than allowing a fairly long silence. Since you must turn away to work at the board, increase the volume of your voice, if necessary, to make sure you can be heard.

Keep the board clean. Completely erase unrelated material and, when your point, word, list, or idea has been satisfactorily covered, clear the board.

Fig. 8-9. Chalk boards/drawing boards are essential tools.

Erasing with long regular strokes will usually leave a cleaner, more uniform, writing surface than you will get by scrubbing with the eraser in short, choppy motions.

"Sugar Chalk"

In addition to using colored chalk to highlight key words or to "color code" a drawing, there are a couple of old tricks you might find useful. One is "sugar chalk."

If you write or draw on the board with chalk that has been soaked in a mixture of sugar and water, the markings cannot be erased. When you are ready to remove the sugar chalk from the board, it washes off easily with water and will not damage the surface of the board. There are a lot of times when this technique is useful—illustrating a form or worksheet which you want to fill out in different ways, for example, or putting up a building layout on which you want to show several different problems in tactics. Use regular chalk to fill in the form or lay out the hose lines; erase the regular chalk when you are ready to move on; the form or the building will remain on the board, good as new.

To prepare sugar chalk, use a wide-mouthed jar or other container deep enough to hold full sticks of chalk, put in two or three ounces of ordinary sugar, fill it with water, and mix it thoroughly. Soak the chalk for at least 20 or 30 minutes, long enough to wet it all the way through, but don't leave it in too long or the chalk will crumble. How long is too long? That depends. We know for certain that overnight is too long. Beyond these highly scientific parameters, you will just have to experiment (Fig. 8-10). The wet chalk will go on the board clear and then will dry to a bright white in five or ten minutes. The technique is useful and it works great after it dries, but it is a

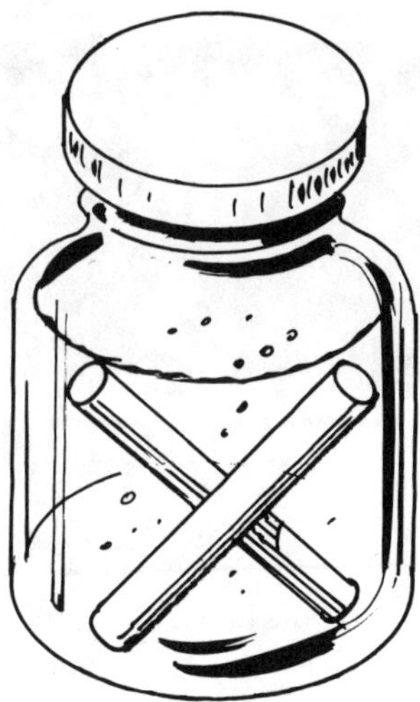

Fig. 8-10. Sugar chalk—messy but effective.

real nuisance to put on the board. The wet chalk is sticky, and it breaks very easily. Give yourself plenty of time to work and have some wet paper towels ready and waiting. If you don't get all that sugar off your hands, your fingers stick together, and you'll draw flies!

Paper Patterns

Sign painters often use paper patterns made with tiny perforations along the outlines they want to transfer to the sign. With the paper taped in place over the sign, they dab the perforated lines with a little cloth sack full of light or dark chalk dust, depending on the color of the background. The chalk dust penetrates the perforations and transfers the outline to the sign. The same technique works very well for putting a drawing or a form on the chalkboard, particularly if it is something you want to use over and over again in different classes, or if it is something in which you want to be particularly neat or accurate (Fig. 8-11).

A dressmaker's pattern wheel, the kind that looks like a little spur on a handle, works well for making the perforations. Lay out your drawing on brown wrapping paper, place the paper on carpeting or some other surface which will allow the wheel to punch through the paper, and run the pattern wheel along the lines you want to use. Flip the paper over and, with very fine sandpaper, carefully knock off the paper points on the reverse side to prevent them from being pushed back and filling up the holes. Tape the pattern to the chalkboard before class, pat the outline with a dirty eraser or a

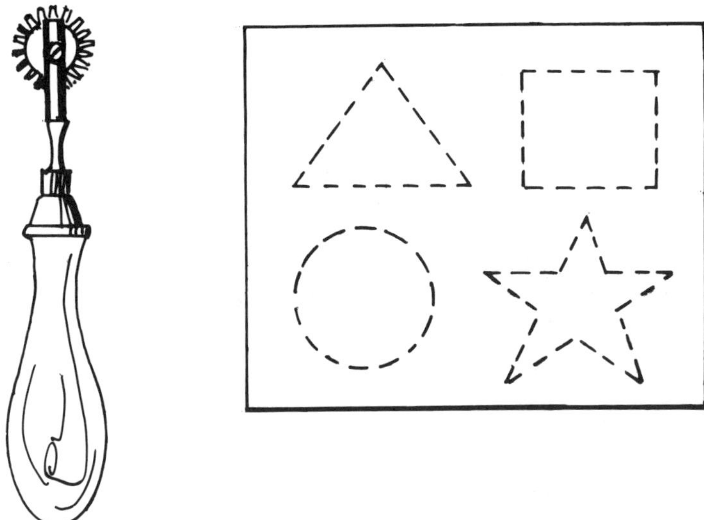

Fig. 8-11. Pattern wheel and paper patterns.

bag of chalk dust, put away the pattern, go over the outline with chalk, and you're all ready.

Charts, Diagrams, Graphs, Maps, and Photographs

If you use any of these graphics at the front of the room, they must be large enough so all your students can see. It may not be necessary for the students to see every detail, but they must be able to follow whatever point you are making. If the material is important, and if you think your students will profit from taking a closer look, post the graphic so they can examine it at break time or after class.

Be aware that if you pass items like these among the class while you are teaching, you will automatically lose a large measure of their attention. Any items you circulate, or any handouts you distribute during class, will immediately pull the eyes and thoughts of your students away from you. If you want their attention on the handout, that's fine. Otherwise, try to avoid giving yourself an unnecessary handicap.

Projections

Projections include all aids which can be projected on a viewing screen. With the advent of video as a teaching tool, it is helpful to subdivide

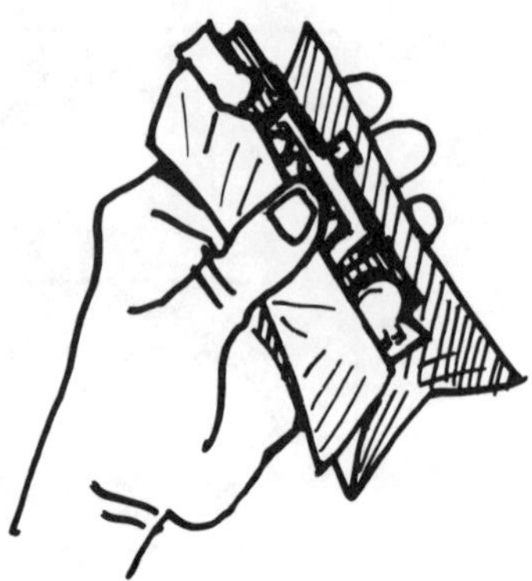

Fig. 8-12. Always protect projection bulbs when handling them.

projections into light-beam projections and electronic projections. Let's first look at some general rules for working with machines which project images via beams of light.

General Rules for Projectors

We have mentioned along the way that we usually prefer to make suggestions rather than to state rules. There is, however, one hard-and-fast rule concerning projectors. Always have a spare bulb. The projection bulb will seldom burn out if you have a spare; it will *always* burn out if you do not. In fact, it's not a bad idea to have a couple of spares—better safe than sorry.

Keep replacement bulbs in their original protective wrappings and cartons until you need them. When changing projector bulbs, do *not* touch with your fingers the glass surface of the new one you are installing. As you place it in the projector, hold the bulb by means of the paper wrapper in which it is packaged (Fig. 8-12). With most high-wattage, high-intensity projection lamps, the slightest fingerprint, caused by the natural oils in your skin, will concentrate enough heat at that point to cause the bulb to fail the instant it is turned on.

We feel strongly that you should not waste class time and create a distraction by floundering around setting up and focusing a projector during class. If the screen and projector cannot be in place before you are ready to use them because of other teaching requirements, set up the equipment before class time, adjust the focus, and mark on the floor with chalk or masking tape the location of the screen legs and the position of the projection stand. If you must place the projector on a table, mark that spot with tape. When you have everything positioned, move the equipment aside to a handy location where it will not be accidentally damaged. When you reach the point in your instruction where you need the screen and projector, you can slip them into place with a minimum of wasted time and motion and with only minor "fine tuning" of positioning and focus. We probably will not ever have the opportunity to evaluate *your* teaching technique, but, on the unlikely chance that we *might*, we suggest that you *never* waste class time while you place a projector or screen by "trial and error." Either follow the procedure we have outlined, or arrange your lesson so the projection set-up can be made during a break. It may be just a detail, but we think it is an important one.

Motion Pictures

Training films, generally 16 millimeter, still pay a big part in many fire service training programs (Fig. 8-13). Your students will profit more from the films you show if, before using any film in class, you preview each one and make some notes:

1. Does this motion picture support the objectives established for this lesson?

2. Are there any points in this film which need emphasis or clarification?

3. Is any of the information shown unsafe, incorrect, or obsolete?

4. At what point will the film best fit into the lesson as you plan to teach it?

Before showing the film, discuss with your class the purpose in showing the film and direct their attention to any particular parts or points you want them to notice. Note any unsafe, incorrect, or obsolete information, assuming the improper information or practice is limited to only a minor part of the total. If there is any significant amount of faulty material, the film should not be used.

We remember sitting in on a class one evening, observing an instructor, when a film was shown in which the protective clothing was obsolete, the equipment was obsolete, and the evolutions were, by current standards, antiquated and unsafe. We went back to the home office and asked why that particular film was still being used by our agency. The answer was that no

Fig. 8-13. Movie projection still provides the brightest image. *Courtesy Eastman Kodak Company.*

one had ever questioned it before—just a case of what was once "state of the art," becoming gradually outdated without anyone noticing. We threw it out. Keep an eye on the quality and accuracy of the films *your* department or agency is using.

Critique the main points of the film after the showing. If it is worthwhile, consider re-running the film, after class or during a lunch break, for those who wish to see it again. Most good training films contain more information than students can absorb in one viewing.

To Show It All, or Not

Because films have a well-defined beginning and ending, we become accustomed to showing the entire reel. If only five or ten minutes of a thirty-minute film directly meet your lesson objectives, *do not hesitate* to show only that part of the film which is really worthwhile. Before class, run it through the projector to the point where you want to start, and when you have shown what *you* want to show, turn it off.

Again, if the group expresses an interest, you can run the entire film outside regular class time for their voluntary viewing.

Similarly do not hesitate to stop a film to emphasize or explain a point. Rewind and repeat a portion as many times as you feel it necessary. A training film is a teaching tool; it is not a time filler, and it is not entertainment. Use it to your best advantage and the best advantage of your students. Incidentally, every one of the points we have made about motion pictures apply equally well to videotapes.

Operating Projectors

We are not going to tell you how to thread film through movie projectors or specifically how to operate *any* projection equipment. Consult the operating instructions or talk with an experienced operator. Just be sure your have *mastered* the operation of your projector *before* class—that is, unless you really enjoy sweating and being all thumbs while twenty self-appointed experts give you their opinions of why the image is jumping all over the screen or why the sound won't work.

Before class, advance the film through the projector to the title or other selected starting point, set the focus, adjust the sound volume; in general, have the film showing *ready*. The goal is to come to the point in class where the film is to be shown with nothing left to do but turn out the room lights and turn on the projector. That's the way it *should* be done.

Fig. 8-14. Slides provide great flexibility. *Courtesy Eastman Kodak Company.*

Slides

The 35-millimeter slide is one of the most popular visual aid tools (Fig. 8-14). In spite of price increases, slides are still relatively inexpensive compared with other media, and slide film and processing are universally available. Slide shows are easy to put together, and they are easy to change and update. If you can shoot slides of good quality on your own, or have access to someone who can take quality slide shots for you, slides are an excellent means for bringing local scenes and local problems into the classroom.

As useful as they are, we want to wave an important caution flag first. In the classroom, slides are often misused. In the chapter on methods of instruction, we mentioned, under illustrated lecture, the fact that entertainment sometimes becomes more important than teaching. Slides have a sneaky way of making that happen. When you see an instructor armed with two 140-slide trays for a fifty-minute lesson, you can bet with absolute certainty that very little teaching is going to occur. There may be *some* learning, and everybody will probably enjoy the hour, but that particular session is going to be long on entertainment and short on education. If you

Fig. 8-15. An illuminated slide sorter is an essential tool.

use slides regularly in the classroom, check yourself from time to time. Are you really using them as a basis for teaching or are you merely dazzling your students with neat and keen pictures of fires and fire apparatus? Like dazzling your students with your skill in tying knots, it may be fun, but "it ain't teaching."

How Many Is Too Many? How Long Is Too Long?

We have no magic number for how many slides are appropriate for a fifty-minute class, nor do we know of any specific length of time a particular slide should be left on the screen. We *are* sure that a particular slide should be shown *only* if it contains information pertinent to the class you are teaching, and we are sure it should be left on the screen as long as it is useful in assisting you to illustrate or explain the points you wish to make (Fig. 8-15). When you complete the discussion of that slide, get it off the screen. Go to the next slide or to a blank. Leaving the same picture on the screen for ten or fifteen minutes, while the flow of teaching has moved on, far beyond the point that slide was used to make, is just as poor a teaching technique as flashing four or five slides a minute—maybe even worse, because it is so deadly boring.

Murphy's Law at Eight to One

There are eight ways to put a slide into a tray or carrier, and seven of them are wrong. That's one good reason for putting slides into trays and keeping them there. The other reason is protection; stored in trays, slides

Fig. 8-16. Marking slide sets saves errors.

are less subject to damage and they will stay cleaner. Unfortunately, most of us cannot afford enough trays to keep all our slides stored in that fashion. If you have sets of slides that you keep in boxes, arrange each set exactly as it should go into the projector, hold the stack together securely with a rubber band, and put a line of color down the top left-hand edge of the stacked slides with a felt-tip pen. While you have the slides in your hand, also run a diagonal line from the upper right corner to the lower left corner across the top of the stack. The diagonal line will assure you that the stack is in the proper sequence. A slide that is out of order will stand out clearly by causing a break in the diagonal line. If the slides are put into a tray or stack loader with the color mark on the top left-hand corner, they will come up on the screen correctly (Fig. 8-16). Simple stuff, but it will help you beat the odds, particularly if you need to load a tray in a hurry. If you simply *must* march to a different drummer, you can reverse all the lefts and rights above. It won't hurt a thing, and if it makes you happy, it makes us happy.

Getting Ready to Use Prepared Slide Sets

We suggested that you preview films before showing them in class. With slide sets provided to you as part of a teaching package, a preview is absolutely essential. The film will speak for itself. *You* must speak for the slides.

Run through the slide set, look over any commentary or teaching points provided, and again make some notes for yourself:

1. Does the slide set support the objectives of the lesson?
2. Does the slide sequence flow smoothly from one point or idea to the next?
3. Can you teach directly from what you see on the screen or will you need to make an outline to avoid missing key points?

4. Do you understand why every slide is in the set, and can you comment intelligently on each one?

You must become completely and thoroughly familiar with a slide sequence before presenting it to your class. With slides, *your* commentary is essential to maximum student learning.

One final thought: If you find a slide that just does not make any sense to you, pull it out of the tray and close up the gap. We feel that is a much better alternative than throwing the slide on the screen, just because it is in the canned sequence, and telling the class you have no idea why it is there or what it is supposed to show. We've seen that situation happen a lot more often than you might expect.

Slide-Tape Shows

Slide-tapes packages are canned teaching programs in which the commentary is provided on tape rather than by the classroom instructor—a low-key form of machine teaching (Fig. 8-17). The concept is fine and we have seen some useful sequences, but, because there is no instructor/student interaction, we feel slide-tape should be limited to a relatively small percentage of class time.

Most contemporary slide-tape machines provide the option of showing the slides on a small screen, internal to the machine, or projecting the images on a conventional classroom screen. It is fairly common to see an instructor misuse the small internal screen with a group of students so large that many of them cannot get a good view of the picture. Make sure that all the students can see.

Establishing proper synchronization of the tape with the slides can be tricky, and it is extremely frustrating for both you and your students when the tape is talking about slide five while slide seven is on the screen. Be absolutely certain you know what you are doing before you attempt to use a slide-tape machine in class.

Overhead Transparencies

Overhead transparencies and overhead projectors may be found in most classrooms. The reason they are so common is that they are excellent teaching tools. The projector optics are designed to throw the projected image behind the instructor as the instructor stands facing the projector stage

Fig. 8-17. Slide-tape projection enhances specialized subjects. *Courtesy Eastman Kodak Company.*

and the class. This arrangement, and the high intensity of the overhead projector light beam, provide some very real and worthwhile advantages.

The high-intensity light and the "transparency" of the materials we use on the projector make it unnecessary to dim the room lighting. Facing the class, with full, or nearly full, room lighting, the instructor can work at the machine and still maintain constant eye contact with the class. The image the instructor sees on the projector stage is exactly the same image students see on the screen; when you point at an item on the stage, you are pointing to that item on the screen.

Setting Up an Overhead Projector

Because we are working between the projection head and the screen, it is essential to arrange the projector and any other items you want to use in such a fashion that you will not need to pass through the projector beam. If you are also using a chalkboard or a table for handouts or other teaching materials, place the lectern with your teaching notes between the overhead projector and the other items you want to use. You can then move back and forth from one to the other as necessary without casting a shadow on the screen or blinding yourself in the process.

In setting up and arranging the overhead projector, there must be enough space *at* the machine for two stacks of transparencies: those you are working your way through, and those you have finished using. It is absolutely worthwhile to take the time and set things up right to begin with (Fig. 8-18). Working with a series of transparencies, making transitions on to and off the projector stage while teaching at the same time, is much more awkward than punching a button on a slide projector control unit. Give yourself plenty of room; you will find you need it.

While we are talking about getting transparencies on and off the stage, we should mention that there are available specially designed overhead projection tables with a cut-out space or "well" in which the body of the projector fits. There is an ample working area on either side of the well at exactly the same level as the projector stage; a superb arrangement for making the necessary transitions with a minimum of lost motion.

It is usually taught that you should turn off the projector lamp each time you remove and replace a transparency. We're not sure. It seems to us, at times, that having the light go off and on repeatedly is almost more distracting than seeing things disappearing and appearing on the screen. It probably is a matter of individual pacing. If you are moving along rapidly, it may be better to leave the light on; if you are working at a more leisurely pace,

 Fire Instructor's Training Guide

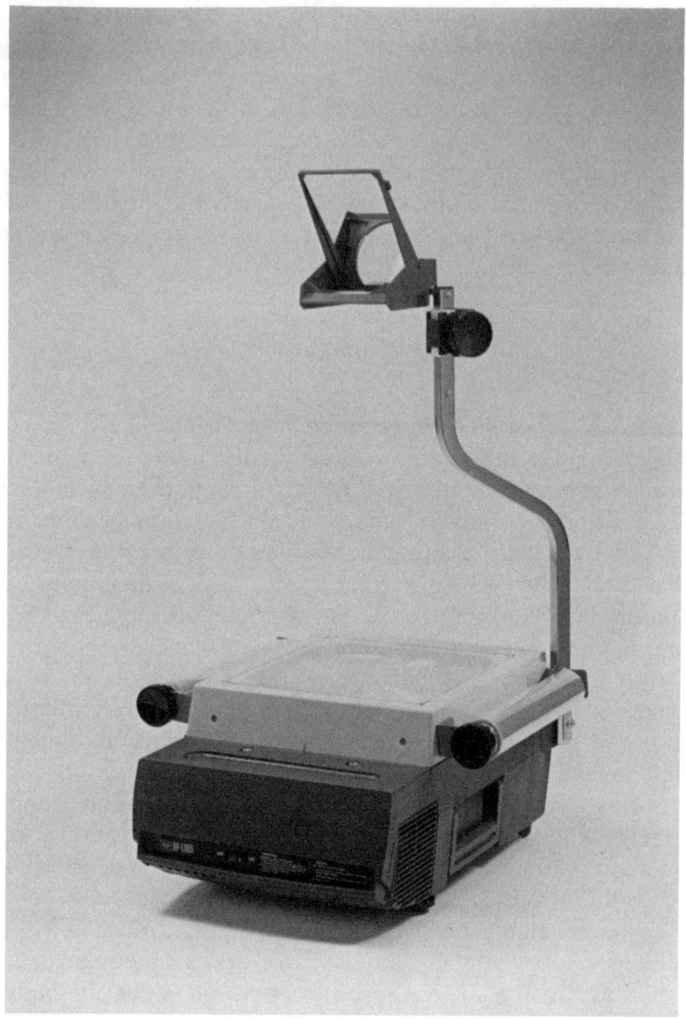

Fig. 8-18. Overheard projectors—excellent teaching tools. *Courtesy Elmo Manufacturing Corporation.*

making a point or two between transparencies, perhaps it is better to turn the machine off. You may want to try both methods.

Keystoning

Keystoning refers to a particular type of projection distortion which occurs when the light beam is projected up on to the screen at a particularly

Fig. 8-19. "Keystoning" should be eliminated or minimized.

steep angle (Fig. 8-19). The image on the screen will be wider at the top than it is at the bottom, assuming the shape of the keystone at the top of an arch. The same type of distortion will occur horizontally, if the light beam strikes the screen at an angle from the right or left.

While most projection arrangements cause some keystoning on the screen, the problem is accentuated with overhead projectors because they must be used closer to the screen. Vertical keystoning can be reduced or eliminated by tilting the top of the screen forward or raising the projector. Many portable screens have "keystone eliminators": an arm which flips forward at the top of the support rod to move the top of the screen toward the projector. Horizontal keystoning can easily be controlled by keeping the projection beam at a right angle to the screen—just another one of those little details which may not seem terribly important, but which separate great instruction from good instruction.

Hints for Going "On Stage"

If you find transparencies skating around on the stage as you work with them, keep a few short strips of masking tape, stuck by one end to the edge of the projector, and use one or two, as needed, to secure the frame of the transparency to the stage.

When you are projecting a list of items, uncover them one at a time, honoring the "singular" concept: discussing each item as you display it. Covering the transparency with a sheet of paper and sliding it down to display one item at a time is called "slipsheeting." If you put the paper on *top* of the transparency, it will often slide off the stage when you get near the bottom of the frame. Put the slipsheet *under* the transparency and it will stay in place all the way down. If you use any lightweight plain paper to slipsheet, you will have the advantage of being able to read the upcoming items on the

Fig. 8-20. Use a small pointer on the projector stage.

projector stage even though they are blanked out on the screen—very helpful in keeping a nice, even flow of information and making smooth transitions from one point to the next.

Give special attention to transparencies with multiple overlay leaves to make sure you flip them over in the right sequence. Check in advance to see if they stack up, one on top of the other, or if you must use them one at a time in rotation. You may find both types in the same set of transparencies. It is almost always worth taking time to secure overlay transparencies with a piece of masking tape; it makes working with them much easier.

You can point to an item on the stage with your finger, but it is better to place a small pointer right on the transparency (Fig. 8-20). A pencil or one of the little arrow-shaped "cocktail picks" works well.

Transparencies

There are some excellent fire service transparencies available commercially. Most come with some form of lesson plan, script, or accompanying training manual. There is certainly nothing wrong with using the transparencies with this type of proprietary instructor's guide, but neither is there anything wrong with changing the order of presentation or breaking up the original sets, integrating them into *your* lesson plans and supplementing the commercial transparencies with those you make yourself.

Most standard office copiers will make transparencies in some form, usually only as a black image on a clear background. Quality ranges from poor to excellent, depending on the machine and the transparency stock used. Thermal copiers make high-resolution, high-quality transparencies and usually give you a much wider range of options in the types of materials you can use. There are different weights and different colors of stock available for both positive and negative images.

Making Your Own Transparencies

If you want to "do it yourself," there are special transparency pens available in most large office supply or school supply stores. Most felt-tip pens will not work well on plastic; many will not work at all. Transparency pens are available in all standard colors, and the inks are designed specifically for creating overhead projections. There are "water-soluble" inks and "permanent" inks. The water-soluble types have the advantage of being easy to erase with a damp cloth and the disadvantage of smearing just as easily if you accidentally brush across them with a sweaty hand. The permanent inks have the advantage of being smear-proof and the disadvantage that they make correcting any errors in your work extremely difficult. Some of the "permanent" inks can be cleaned off with a mild alcohol-and-water solution, but this solution may or may not etch the surface of the transparency stock and, at best, it is difficult to control.

Grease pencils or "china markers" were used on transparencies for years, before the special pens were available. They are troublesome to work with because they break easily, they will not hold a fine point, and they project only a black image on the screen. While they are certainly not the best choice, they still work and might help you out if you find you need something to draw or write with but none of the special pens are available.

If you have an artistic talent, you can create your own work, and, if you don't, you can always trace. As with making graphics, you are limited only by your imagination and the time you can devote to preparations. The overhead projector is an extremely versatile tool. There are many different "add-ons" on the market: motor-driven polarized wheels, for example, for use with polarized films which can be applied to transparencies for creating motion— excellent for showing flow through pumps or any other fluid actions. We remember once seeing especially designed clear plastic trays which, when filled with colored water and placed on an overhead stage, were used to project the image of actual fluid flows, demonstrating the operation of transfer valves and similar devices.

For creating your own quality transparencies with wording rather than art work, there are many options. You can hand letter, obviously, and many a class has been taught using home-made, hand-lettered overheads. If you have available a typewriter with extra-large type, like one of the old "Western Union" machines or the modern "orators" ball, you can type out paper masters and copy them as transparencies. For a more professional quality,

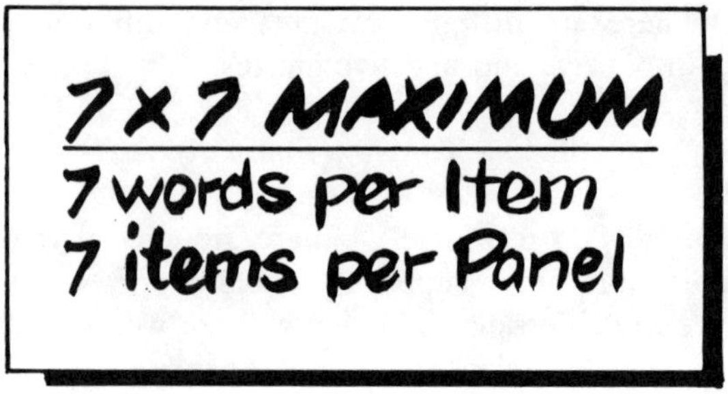

Fig. 8-21. A good rule of thumb.

and at considerably more expense, there are special systems for "typing" your own lettering in a wide variety of type faces and sizes on special self-adhesive transparent tape. The finished wording is cut as appropriate, peeled from the backing material, and arranged as you want it on a layout sheet. The finished product is again copied on to overhead stock by one of the methods described earlier.

Wording on overhead transparencies should be limited. One rule of thumb states that there should be no more than seven items per panel and no more than seven words per item (Fig. 8-21). We suggest you use it only as a guideline, certainly not as an absolute restriction.

It is sometimes suggested that overheads should be arranged *only* in a horizontal format. This rule makes sense for those with wording, but we disagree where drawings or other illustrations are involved. For art work, a vertical concept should not be forced into a horizontal layout.

Opaque Projectors

We debated over whether or not we should mention opaque projectors at all. As the name implies, they are designed to project opaque, or non-transparent, materials. They will also project the image of any object small enough to fit into the projection tray or slot. Inside the machine, the inserted graphic or object is flooded with light, and the reflected image is optically projected on to the viewing screen. Compared with all other light-beam projectors, the image produced by the opaque projector is quite small and lacking in brilliance. Unfortunately, the heat from the lamp or lamps is so intense that it is possible to scorch or actually ignite a flammable material, like paper, which is put into the machine.

Before the ability to create transparencies became almost universal, in even the smallest educational offices, the opaque projector was our only readily available alternative for getting some image on the screen without going through a lengthy and cumbersome photographic process. It was also the most convenient way to enlarge and copy graphic materials, for example, projecting a small drawing on to poster board to create a large graphic by tracing and filling in the outlines—exactly the same process we described in the graphics section of this chapter.

There may very well be an occasional specialized application for the opaque projector, but in most fire training agencies they are seldom used.

Projection Screens

Projection screens come in different types and sizes and are constructed with different reflecting surfaces (Figs. 8-22 and 8-23). Large, wall-mounted screens are the preferred arrangement for permanent classroom facilities. For most classrooms of ordinary dimensions, a good-sized manual pull-down screen will most probably be big enough. In large classrooms and major meeting rooms, a motor-driven screen is preferable. IFSTA recommends the audience should be seated no closer than twice the width of the screen and not farther away than six times the screen width. These numbers will help you in selecting a room screen of appropriate size.

Folding tripod-leg screens are available over a wide range of sizes and have the obvious advantage of portability. They are, however, easily damaged, easily soiled, and, in the larger sizes, can be heavy and difficult to handle. If portable screens are used and frequently moved in your training agency, zippered protective covers are a must. When these wear out, replace them. It's a lot cheaper to buy new covers than it is to buy new screens.

We prefer glass-beaded portable screens because of the high reflectivity and bright image they provide. We are not going to go beyond that into all of the other various screen surfaces which are available. If you are making a major purchase of portable screens or a large room screen, discuss the surface options with a number of manufacturers or suppliers and visit or contact one or two educational agencies which have purchased and are using the type you select, *before* you sign a contract. It may save you considerable grief.

We do not care for rear projection screens unless there is, as with the simulator, a compelling reason for wanting to get the projection equipment out of the student area. Rear projection screens are relatively expensive, highly susceptible to damage, and they do not, in our opinion, provide as brilliant an image as surface projection screens.

Fig. 8-22. Typical portable projection screen. *Courtesy Da-Lite Screen Company, Inc.*

Video and Video Cassettes

The second form of projection is electronic. Video and video cassettes are playing a constantly increasing role in training and education. The quality of commercially produced fire service training video cassettes is improving each year (Fig. 8-24). Many, though, are too generic to have much value in a well-structured training program. Choose your materials carefully

Fig. 8-23. Typical classroom screen installation. *Courtesy Da-Lite Screen Company, Inc.*

and take note of the points we made relative to motion pictures. Those thoughts are, as we said earlier, applicable to video.

The tremendous advantage of video recording is the immediacy it provides: shoot it, rewind it, and play it back. For us in instructor training, there is no more vivid experience we can give novice instructors than to tape their practice teaching sessions and let them see and hear for themselves exactly how they look and sound from the student's viewpoint. The same advantage extends to any form of skill training: physical skills on the drill ground or administrative skills such as the television "interview" at Dorking we described earlier.

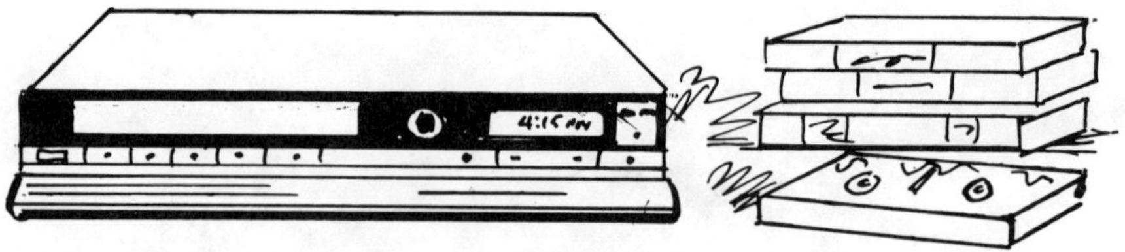

Fig. 8-24. Video—growing daily.

The training division of any fire department should also take an active role in shooting videotape of actual emergency operations whenever possible. Playback analysis of fireground evolutions is an excellent tool for determining how well your standard operating procedures are being followed and for detecting any weaknesses in the training program.

Our major current concern about video is the poor way that most electronic projections are employed in the classroom. Everybody knows that movies are shown on *big* screens and television is viewed on little screens. Because everybody *knows* that, we put two or three 19-inch monitors in a classroom seating a hundred people and sit back satisfied. Can you imagine the criticism you would get if you set up a movie projector and tried to show a film on a screen 19 inches across the diagonal? You'd be chased right out of your classroom.

The quality of the image on the super-large reflected-image television screens used in some classrooms is getting better with each new generation of equipment, but such screens are still a long way from being as vivid as small-screen television or motion pictures. If your training budget can absorb the cost, they provide one answer. The only alternative is a sufficient number of smaller monitors to assure a good view for all students, and that's not inexpensive, either.

Three-Dimensional Training Aids

Graphics and projections have, obviously, only two dimensions, height and width. The third and last group of training aids we mentioned were those with three dimensions. These devices include an almost infinite variety of actual equipment, objects, models, mock-ups, and cutaways.

The three-dimensional aids most frequently used are pieces of actual equipment, the tools of the trade. In most cases, the item or object under discussion will be the best training aid we can use. There are cases, though, when a model or a mock-up may be preferable.

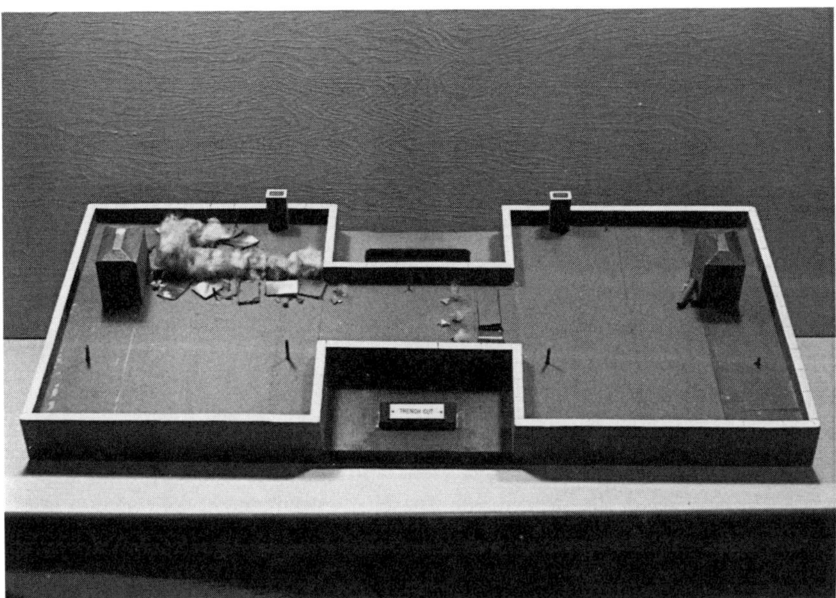

Fig. 8-25. Models fill many classroom needs. *W. N. Carey, Jr.; courtesy Maryland Fire and Rescue Institute.*

Models can be scaled-down or scaled-up versions of real objects (Fig. 8-25). If the real object is too small to be seen easily by the class, or too large to be comprehended readily, a model may be very helpful. We usually think of models as being small, scaled-down versions of the real thing. One example might be a model of a typical flammable liquid storage bulk plant, or perhaps a model of one of the new liquified natural gas supertankers. On these models, the instructor can point out features and details which might be lost in the immensity of the real thing.

Back in the days before hand-held calculators, many engineering classrooms had a huge working model of a slide rule, perhaps six or eight feet long, hanging in the front, right over the blackboard. A standard slide rule would be far too small for students even in the front row to see, so instructors demonstrated methods for working slide rule problems on the scaled-up model which could be seen clearly from the distant seats in the room.

Models are used when real objects are dangerous, like dynamite and dynamite caps. It is useful for fire investigators, for example, to see and handle various types of explosives and blasting caps, but it does not make

Fig. 8-26. Cutaways clarify internal parts.

much sense to pass such extremely dangerous items around a classroom. Models, realistic in size, shape, weight, and finish, solve the problem.

Models may be helpful in demonstrating mechanical actions where the actual movement is too fast or too slow to be readily visible—tripping of an exhauster or an accelerator, for example. Models can also be used to illustrate a phenomenon which is naturally invisible, such as the models of molecules which help explain how elements are bonded together.

Mock-ups are often used to replicate equipment which would be too expensive or too susceptible to damage to use in training. A mock-up of an aircraft cockpit illustrating the hazards and safe practices concerning a rescue involving an ejection seat is an example.

Cutaways are often used to show the position and movement of normally invisible internal parts (Fig. 8-26). Cutaway fire pumps and fire hydrants are found in almost every fire training academy. Color coding is helpful in pointing out specific parts and actions to students.

Finally, be alert for any salvaged or junk materials that can be used as an aid in teaching. Short sections of hose, broken parts illustrating misuse of equipment, articles which can be used to demonstrate a mechanical principle—any item which might help your students learn or understand can be an asset in your personal inventory.

Working with Three-Dimensional Aids

There has been an ongoing argument in fire service training, for as long as we can remember, over whether or not three-dimensional training aids should be openly displayed at the front of the classroom or "hidden away" until they are put into use. We can see some logic on both sides, but we tend toward putting them out where everyone can see them. It is easier, and we feel it provides an element of interest which overshadows the disadvantages of distraction and possible questions from students before you are ready to teach the functioning of a particular item. Both problems can be controlled.

You know, we've just about covered everything that we planned to cover concerning teaching. When we were discussing lesson planning, we said we would come back to the examination process and talk about it separately. Let's get a couple of cans of soda and go outside for a little fresh air. We can talk about testing and evaluation while we stretch our legs a bit.

9

Testing and Evaluation

Evaluation of student progress at various stages within a course of instruction is a fundamental part of any educational system. In the highly competitive world of the fire service, the bottom line, usually, is who passed, who failed, and what was the rank order of those who finished the program. Testing and evaluation, however, can and should tell us a great deal more than just who scored what.

We should, ideally, monitor the progress of each class, or section, as a unit, measuring the success or failure of the entire group against the established criteria and against the known performance of previous classes. We will certainly want to evaluate instructor effectiveness, and group performance is the key element in determining how well any instructor has taught. We must continually monitor the accuracy and the validity of the evaluation instruments we are using. Finally, it would be advantageous, if we can, to assess the effectiveness of the learning environment. Obviously, it is not an easy task to isolate any one of these elements cleanly and clearly.

In a large system with many students, many classes, and many instructors, we can use comparative analysis carefully and cautiously to make judgments from course to course, from instructor to instructor, from test to

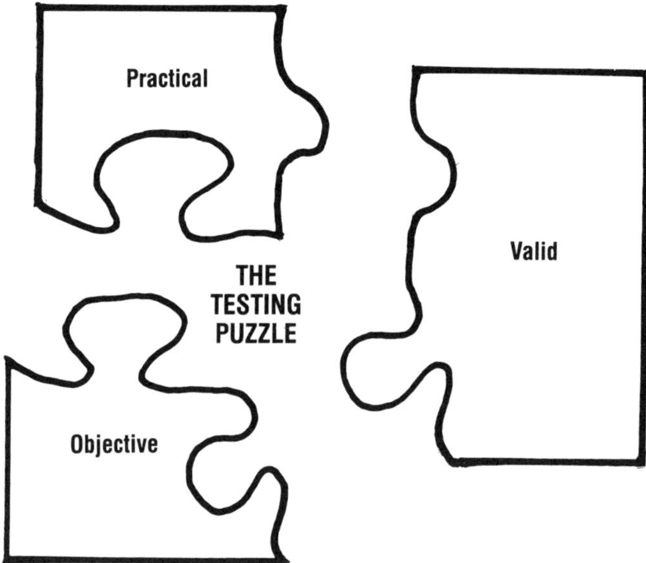

Fig. 9-1. Putting the pieces together.

test, and, possibly, from environment to environment. Large and varied statistical bases provide norms against which we can measure.

The smaller the system, the more difficult it is to make these peripheral evaluations accurately. With one instructor, teaching one group of students, in one environment, measuring the results with one test, it is extremely difficult to determine the relative effect of any single element on the total success or failure of the effort.

Questions and Tests

With respect to specific questions and tests, which are nothing more than groups of questions, we should make sure, to the degree that we are able, that these are valid, objective, and practical (Fig. 9-1). A question has validity if it does in fact measure what it is supposed to measure. A question is objective if personal opinions and subjective judgments are eliminated or controlled during grading. Practicality is more a measurement of tests than of individual questions. A test may be judged practical if it is easy to administer, easy to score, and economical to use. Objectivity and practicality are

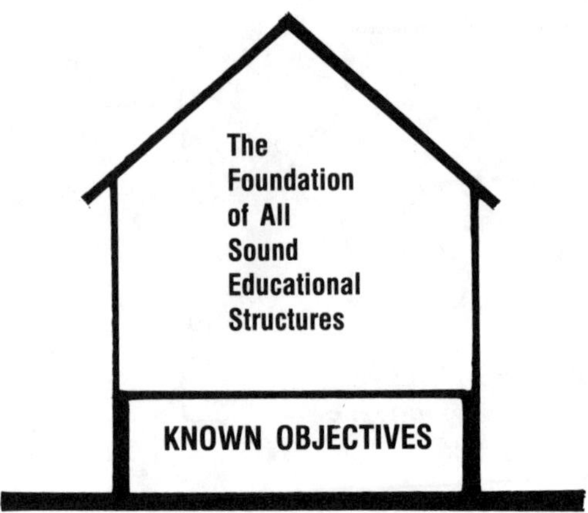

Fig. 9-2. Educational credibility.

easily observed and can be judged with common sense. Validity is much more elusive, harder to determine.

Before we close this chapter, we will discuss the evaluation of test scores and the kind of information which we can gain by analyzing test results. First, however, let's consider the central problem: evaluating student progress.

The Importance of Performance Objectives

Properly written objectives provide us with a known and solid base for testing. That's one of the reasons objectives are so important and one of the reasons we spent so much time discussing them. The entire sequence of testing and evaluation for any course, from quizzes through the final written and practical tests, *must* measure student ability to achieve the *stated objectives. Everything else is out of bounds.* If you want to bring an "out of bounds" item into the testing procedure, you *must* write it into the course objectives and make it known to the participants.

Performance objectives, when they are handed to each student and when they form the core of the instructional outline, give our examination process both educational credibility and legal validity (Fig. 9-2). In the world in which we live and teach, both are absolutely essential.

Informal Evaluation

We evaluate informally, as well as formally, in every class we teach. Your informal evaluation begins in any course when the students begin to drift in the door for the first class session. You cannot avoid mentally classifying those you have taught in previous classes. The appearance of known names on the roster, or the appearance of certain individuals in the classroom doorway, will create in you some form of reaction: positive, negative, or neutral. Those you do not know will produce some form of influence on the recesses of your mind through that all-important "first impression": appearance, the manner in which they greet you, or just the way each enters the room will be mentally recorded and filed away, even if you try to avoid doing so. None of us is immune to these almost subliminal judgments. The key is to recognize them and, to the best of our ability, to control the effect that they have on our official evaluations.

From the opening session onward, you will continuously gauge and note down the varying performance of each of your students. Throughout any course, you will informally evaluate student knowledge as you pose questions and mentally measure the accuracy of student answers. Your observation of student abilities during practical evolutions provides a similar opportunity to evaluate informally individual student progress in skills. Well before the final examination, a good instructor will be able to estimate, with a fair amount of precision, who will pass and who will not, who will rank high, and who will be at or near the bottom of the list.

Informal evaluation will also give you an idea of your personal effectiveness. A high general level of student comprehension and skill is an indication that your teaching is going well. A low general level of student comprehension and skill is a warning that something is wrong. As we suggested earlier, the first place to look, when you sense this kind of problem, is on *your* side of the lectern.

If the class is generally doing well, your informal evaluation will help you to identify individual students who may need special assistance or encouragement. The earlier you can determine that a student is drifting into deep water, the easier it should be to pull that person back to safety.

Formal Evaluation

Formal evaluation is characterized by recorded scores or grades which students achieve in written quizzes or exams or in practical performance testing. We encourage you to consider using periodic quizzes during any lengthy course of instruction. Quizzes familiarize students with the type and style of questions they will be expected to answer, and the quizzes indicate to students that you are totally serious about the importance of the course objectives. Quiz scores can be recorded and made a part of the complete course grade, or they can be viewed merely as diagnostic tools which give the instructor and the students an idea of how the work is progressing.

We will, with some hesitation, make some general suggestions for preparing both written and practical tests. The reason we hesitate is that this is such an important area, and it is extremely difficult to assess the validity and value of any particular quesiton. We urge you to build your library of written test questions and practical examination procedures with care. Keep an exceedingly critical eye on the construction, results, and utility of every individual test item.

The impact of the examination process on student progress and on student careers is obvious. Unfortunately, in our experience, test preparation is too often treated as the least important aspect of course development, an unpleasant and uninteresting afterthought which must be gotten out of the way. All of us in training and education should move carefully, almost humbly, when we are developing any evaluation instrument. If there is any part of our profession which demands our best attention and a total absence of professional arrogance, this aspect is the one.

Written Test Questions

There are a variety of types of questions we can use in written quizzes and examinations. The most common types are multiple-choice, true-false, matching, completion, and essay (Fig. 9-3). It is possible to construct excellent, searching, valid questions using any of these types.

Unfortunately, it is also possible to construct poor questions: questions which have more than one correct answer, questions which can be interpreted in an entirely different way from that which was intended, questions which unintentionally twist meanings, and questions which answer other questions. *You* will not find your bad questions. Not to worry—your *students* will find your bad questions, and they will let you know *immediately* when they find them. Except for questions which answer other questions—for some reason, students never complain about those. (At least, ours don't.)

If you discover there are one or two bad questions on a lengthy test which has already been reproduced in quantity, you can create a proper question and insert a copy in each test booklet, simply by pasting over the offending question. If test keys, manual or machine, have already been made, be sure the new question has the same answer response code as the original. Short tests or quizzes should be rewritten and rerun to eliminate any poor question.

In the measurement of temperature level, 32 degrees Fahrenheit is equal to 0 degrees Celsius.

True-False Question

In the measurement of temperature level, 32 degrees Fahrenheit is equal to:
  a. -10 degrees Celsius
  b.   0 degrees Celsius
  c. $+10$ degrees Celsius
  d. $+22$ degrees Celsius

Multiple-Choice Question

In the measurement of temperature level, 32 degrees Fahrenheit is equal to _______ degrees Celsius

Completion Question

Briefly discuss the relationship between the Fahrenheit and Celsius scales in the measurement of temperature level.

Essay Question

Fig. 9-3. Test questions.

There is a "quick and dirty" solution which we don't recommend for continued use but which can rescue you if you must use a test before it can be corrected or redone. Merely tell students to "Ignore question 7," or tell them, "Question number 7 is a bad question; mark answer B for a ———- point bonus." Just make sure students know *they* will not be penalized because *you* let a bad question slip into the test.

Let's leave this unpleasant, negative subject of how to handle "bad" questions, and talk positively about how to write "good" ones.

Multiple-Choice Questions

Multiple-choice questions have a great many advantages. They are thoroughly familiar to most students, which makes them a relatively "comfortable" testing tool. They are easy to score—an important characteristic when many examinations must be handled. Multiple-choice items are also quite versatile. They can be arranged in a variety of different patterns, giving you a considerable flexibility without violating the multiple-choice format.

Multiple-choice questions are constructed in two parts: the stem, which sets up the problem, and the responses, from which students select an answer.

In constructing the stem, make the problem plain and clear. Use words which students can understand and keep the sentence structure as simple as possible. We particularly dislike double negatives, which require the student to "turn the question inside out" before it becomes understandable. Keep the wording direct. We should be testing their fire service knowledge, not their skill in logic or deductive reasoning.

The stem can be written as a question or as an incomplete statement; either is proper, and both work well. It is also possible to use a diagram or drawing as a part of the stem; a technique which permits the use of multiple-choice questions involving objects, mechanical actions, tactics, and many other areas.

Four responses are usually provided, although there is no reason you cannot use three or five if there is a logical reason to do so, and the scoring sheet permits. There must be, obviously, only one correct answer. This requirement is where trouble, most commonly, creeps in. We write, review, examine, read over, agonize, and finally decide a question is foolproof, so we use it, congratulating ourselves on being so clever. We are absolutely sure answer "B" is the *only* possible correct solution. The very first time the test is used and scored, some student who was marked wrong on that question will say, "Yes, but . . . ," and point out that under *this* set of circumstances, answer "C" is also correct. The worst part of this whole scenario is that the student is usually right. Nobody said this was going to be easy.

The incorrect answers are called distractors. They must sound plausible and create the impression that they *could* be correct. It is not easy to write distractors which "look like" possibly correct answers, yet are not ambiguous, not too easy to eliminate, and will not permit the student to "back into" the correct answer on the basis of common sense. The responses should be arranged in some logical order so the right answer does not stand out. An alphabetical series is fine, if that is possible. When the correct answer is a number—200, for example—use a series of regular steps: something like 50, 150, 200, 250 or 190, 200, 210, 220, if that series would be more appropriate.

All responses, the correct answer *and* the distractors, should be approximately the same length. This result is not always possible, but it is a goal to aim for. If you end up with some distractors of irregular length, which are too good to pass up, write two long responses and two short ones and arrange them: long, short, long, short. Each response should be complete, and each must agree grammatically with the stem. It is just as easy to write well-crafted, proper responses as it is to write careless ones. A little extra time and effort will give your examination added polish and credibility. We feel multiple-choice questions are an excellent tool for constructing effective short-answer tests; develop your ability to write good ones.

True-False Questions

True-false questions are also familiar to students. They are easy to score and they require, compared with multiple-choice questions, a little less work to create. On the other hand, they are not as flexible as multiple-choice questions because of their narrow "yes-no" framework.

True-false questions are more likely to encourage guessing, and we believe that, in this respect, they favor poor students at the expense of students who study, particularly in the short quiz format. In a ten-question true-false quiz, a student who knows only five answers for certain, and simply guesses at the other five, has a better than fifty-fifty chance to pass, assuming 70 percent as a passing grade. In a ten-question multiple-choice quiz, with four responses per question, the student who simply guesses five answers faces odds favoring failure at something close to four to one.

True-false questions must be written in extremely precise terms to avoid ambiguity. It can be difficult to write statements which are completely true or false without resorting to absolutes like "always" and "never." Absolutes should be avoided if possible because chance heavily favors an "always" or "never" true-false question as being false. If you write, "Two and one-half-inch fire hose is *always* made up in fifty-foot lengths," I don't have to look for a hundred-foot length to prove the question false. A length of 49 feet 11½ inches or a length of 50 feet 1 inch will do very nicely, thank you. Avoid, also, true-false questions which use comparative terms such as "is better than." These terms only invite arguments on opinions, arguments which waste time and are almost impossible to win.

Keep true-false statements relatively brief, and limit each to only one central principle. Carefully written, these can be used to measure student ability to recall knowledge and, to a lesser degree, to apply knowledge. Our main concern goes back to our earlier statements concerning guessing. The reliability of true-false questions is highly questionable unless they are used in relatively large numbers, over a wide spectrum of students, and the utility of *each* question is very carefully checked by analyzing the test results. Easier to write, yes, but true-false questions may not give you results which are accurate and fair. Be cautious.

"Short-Answer" Tests

Multiple-choice and true-false questions are those used most commonly on "short-answer" tests in which the student is required only to check off, in some manner, the chosen answer. This fact makes these two types of questions ideal for constructing tests which can be rapidly and easily scored by hand, with a template, or by machine, if they are answered on specially designed, machine-readable forms (Fig. 9-4).

The simple one-page quiz, which is usually scored by the class instructor and returned to the students for their information, can be a "throwaway" item. For short-answer examinations of substantial length, it is much more economical to prepare reusable test booklets and have the students indicate their answers on a separate scoring sheet. This method has the dual advantages of preserving the test booklet and creating a separate answer sheet which is much easier to handle and score.

Opposite the number of each question, the scoring sheet has a pair of lines or a circle for each possible answer. The student selects one answer and blackens the area between the lines or within the circle. If machine scoring is used, forms must be suitable for electronic scanning and soft graphite pencils used to mark the selected answers (Fig. 9-5). Electronic scoring has been generally limited to relatively large-scale, high-volume educational settings, but it gives us some huge advantages. Tied into an effective computer program, this method of grading can provide an almost unlimited range of analytical and administrative options. It should be explored if grading becomes a roadblock in your system.

In a small operation, you can prepare the answer forms "in house," run them on a copier, and make a hand-scoring template by punching holes in one copy at the spots where the correct answers should appear. Grading is accomplished by placing the template over each answer sheet and making a mark of contrasting color, red usually, in every hole where the proper space has not been blackened. Grading is reduced to a simple mechanical task. Total red marks equal total wrong. Limited test analysis can be done by hand, of course, but today even smaller jurisdictions usually have enough computer capability and expertise to enter student scores into some simple computer-driven record and analysis system.

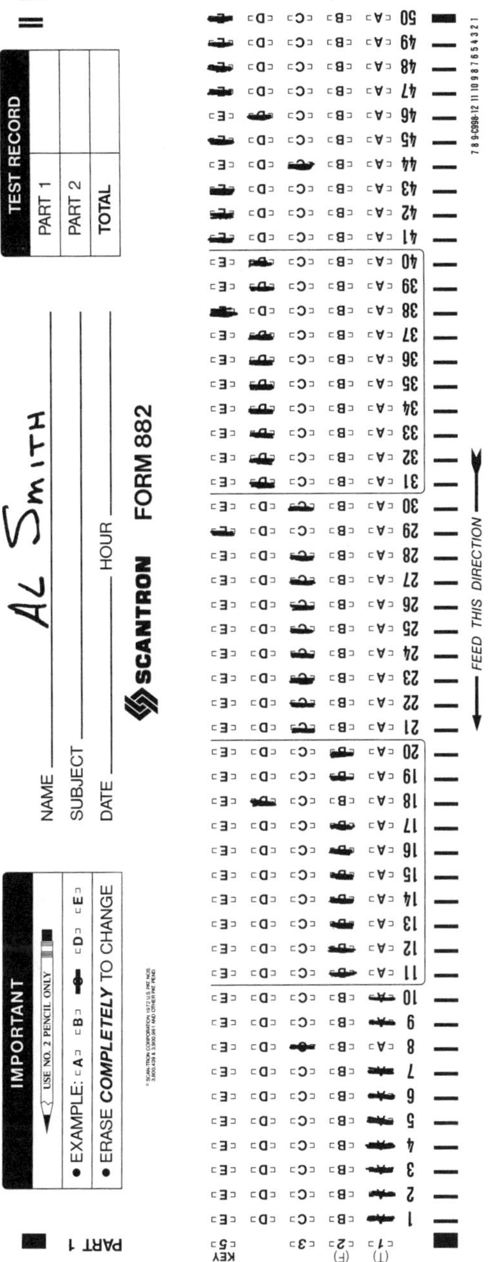

Fig. 9-4. Typical machine grading form. *Courtesy Scantron Corporation.*

233

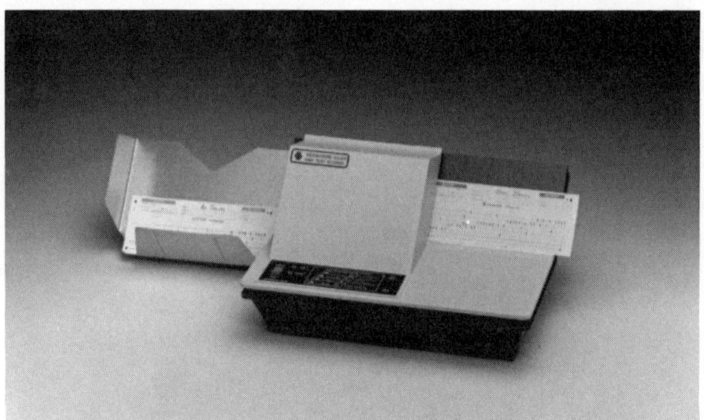

Fig. 9-5. Grading machine. *Courtesy Scantron Corporation.*

Matching Questions

Continuing with types of questions, we touch on matching items primarily because they are fun to construct and fun to answer. From a practical standpoint, they are awkward to score, and we are not convinced that they have a great deal of power to discriminate between good and poor students. They measure only student ability to associate one item with another and may have their principal value in keeping students on their toes, checking to see whether or not they have read assigned material, for example.

Matching questions are arranged with two sets of items, one list of problem items, usually in a vertical column on the left, and a second list of answer items in a vertical column on the right. All of the items in each list should be homogeneous—of the same type or class. If you have a choice, the problem items should be the longer ones and the answer items the shorter. A satisfactory number of problem items is approximately ten to fifteen, and it is generally thought that there should be more answers than problems in order to increase the odds against guessing the last few items.

Items in the problem column should be arranged in a random order and items in the answer column should be listed in some logical fashion—alphabetical order, for example. Scoring is easier if problems are numbered and answers are assigned a letter. Put a blank after each number where the student will write the letter of the answer selected. Unless you want to spend a very long, unhappy evening, untangling knotty problems, *don't* have the students connect matching items by drawing lines between their choices. For the sake of all concerned, keep the whole business on one side of one page. As always, clear instructions and rules on how the game is to be played will save you a lot of trouble. Have fun!

Completion Questions

The final two types of questions, completion and essay, require students to write out answers and the instructor to carefully read, interpret, and grade each answer on an individual basis. This process is obviously time-consuming and exasperating, but it is the only way to reach the higher cognitive areas we discussed earlier in levels of learning.

Completion questions measure student ability to *produce* answers, not merely to recognize them. Clearly, this fact increases our ability to discriminate better because it virtually eliminates guesswork. Subjectivity definitely becomes a factor in your scoring, but it can be kept within reasonable limits if the questions are carefully written to call for specific answers.

In constructing completion questions, it is better to insert blanks only where truly key words or concepts fit, and, if possible, to arrange the sentence structure so the blanks fall at or near the end of the statement. Don't get carried away. Consider a question reading: "The ——— of ——— is ———." Doesn't cut a student much slack, does it? Be reasonable.

Essay Questions

Essay questions require written answers to relatively complex questions, usually under the pressure of limited time. They force students to organize their thoughts, to apply learned principles and concepts selectively, and to communicate their proposed solutions effectively.

Essay questions are the most difficult to score with a high degree of objectivity, but they are the *only* cognitive questions which give the better students an opportunity to show us what they really can do. They are essential tools for evaluating student abilities in areas where advanced application, analysis, and creativity are involved. With the appallingly low levels of literacy we face today, the essay question probably must be limited to relatively advanced programs where a certain amount of self-selection has already winnowed out those who do not have at least some minimal writing skills.

Just as an aside, we find this semi-literacy a terribly frightening aspect of contemporary society. Where is this nation heading, where can it *go*, when we give high school diplomas to people who are, for all practical purposes, functional illiterates, and we award college degrees to people who cannot write an acceptable, simple, business letter? Unbelievable! Be that as it may, the essay question is a legitimate examination device, and it has a valid place in fire service education and training. Use it. Period. End of discussion.

Essay questions should not be used to test simple recall. There are, as we have previously stated, better and easier ways to test that kind of elementary knowledge. Essay questions should require students to display ability to put their knowledge to work—which does not imply that students should be permitted total freedom in responding. Questions should clearly and simply establish the specific area you want addressed, providing as much detail as you feel is necessary to keep student responses on track. It may also be useful to indicate a suggested maximum time or length for each separate essay item on the test.

Since objective scoring is difficult, as you construct the questions you may want to give yourself the advantage of predetermining a specific number of points you will allow for each major factor or concept you expect to find in an acceptable answer. It may not be possible to follow such a hard line exactly, but it will structure your grading and keep you from wandering too far off the mark, particularly if you have a large number of essay tests to score. Without some sort of grading plan, it becomes extremely difficult to remember how you graded the first test when you look with glassy eyes at the twentieth or twenty-fifth.

Essay tests with only three or four questions have an inherent danger, from the student's viewpoint. Since each question counts for a third or a fourth, respectively, of the total score, an incorrect assumption or an incorrect perspective of what one question is seeking can cause the student to lose a damaging number of points. In a senior-level examination you may *want* to assess the ability of the student to interpret and respond correctly to a limited number of highly complex questions. In less intensive essay testing, it may be a better arrangement to increase the number of questions and reduce the length of the responses required. Another useful technique is to vary the length of the responses, including several questions which require only a few sentences, and one or two which require more extensive answers. Tell students, in your instructions, the percentage weight or points you are assigning to each item or indicate this information on the test paper for each question.

One More Time, The Key Word Is———?

Objectives? Right! We have talked about five different types of questions, each with advantages and disadvantages. The common theme which runs through all of them is that they *must* center on the stated objectives you have established. The questions you write may be clever or they may be straightforward. They may prove to be valuable as discriminators or they may prove relatively worthless. They may turn out to be "good" questions or "bad" questions. All of these things we can justify and explain. There is no explanation for a question which asks the student for information which is not clearly spelled out in the performance objectives written for the course.

Written Test Arrangement

In preparing written tests, particularly lengthy major examinations, the arrangement of questions must be considered. Questions can be in random order, arranged in increasing order of difficulty, or arranged to coincide with the order in which the material was presented in class. All three arrangements are commonly used.

We are currently leaning toward order of presentation. It permits students to deal with the points raised by the examination in the same order in which they were studied, and it minimizes the need to jump mentally from one unrelated topic or section of the course to another. It is often possible to blend this approach with the concept of including a few easier questions first to let students "ease into" the exam and inserting a few harder questions near the end to enhance discrimination.

It is useful to note on the exam, next to each question, the source from which it was drawn. You can key the source to the appropriate page in the student material or to the number of a specific enabling objective. This device assists the instructor in any post-exam critique, and it confronts, head on, student complaints of, "Where did they get *that* one?"

Written Test Administration

Many of the things we talked about in classroom arrangement and design become doubly important during testing. The evaluation environment should minimize student stress, not contribute to it. In colloquial terms, the setting should be as "hassle-free" as we can make it. Seating should be comfortable, and table space should be ample. Lighting, temperature, and ventilation levels should all be adequate. Freedom from distraction is particularly important; sights and sounds which attract student attention must be kept to an absolute minimum.

To maintain this freedom from distraction, we have always required students who finish early to leave the testing area quietly. If you allow them to stay in the room, the post-test release from tension and their interest in discussing the test among themselves will make it impossible to control the level of conversation. Our announced rule is, "The last person finished has just as much right to peace and quiet as the first person finished."

It is good practice to remind the students to read the questions carefully, making sure they know what each question is asking. In most cases, we will assist students in interpreting questions if they ask for help, by rephrasing or restating the question in slightly different terms, or giving them an example of the concept involved. This practice may or may not be advisable or permissible, depending on how closely the examination is tied to career advancement. Under certain circumstances, the only proper course of action may be to read the examination instructions, pass out the papers, move to the back of the room where you can observe, and provide no assistance whatsoever.

Unless the examination has a time limit, we usually advise students to take their time: "There is no bonus for finishing first, and there is no penalty for being last." In routine fire service testing, we feel that just about as many points are lost by carelessly reading the questions and by hurrying as are lost through a lack of study.

"Academic Dishonesty" (Formerly Known as Cheating)

Cheating, in our experience, is not as much of a problem in fire service classes as it may be in other educational systems. Unfortunately, it does happen. Every fire training program should have a clearly established and well-publicized policy that cheating will not be tolerated. We feel the best tactic is to arrange seating and table space so that it is impossible for any student to see another's test paper. It may require a little extra work in setting up the room, but it definitely does take much of the tension out of test administration. Furthermore, keep the tables clear of any papers, books, or written materials except those which you distribute and collect during the testing. These steps, plus close and constant, but unobtrusive, monitoring, will keep the students, and you, out of trouble. An old saying fits: "An ounce of prevention is worth a pound of cure." We trust our students, but we watch them like a hawk and we let them *know* we are watching, keeping the whole situation, at the same time, light-handed and low-key.

Examination Document Security

Prevention, again, is the best way to ensure the security of examination answer sheets, answer keys, and all related items. Keep them under close and tight control at all times (Fig. 9-6). Count examination booklets before

Fig. 9-6. Security is essential.

they are issued, and count them again when they are turned in. Putting a serial number on each reusable examination is an excellent idea, as is requiring students to record that examination serial number on their answer sheets. It certainly reduces the number of tests that walk away. Put student answer sheets or written tests into an envelope and seal it until the time comes for grading. Keep the answer sheets, written tests, and all answer keys in secure and locked cabinets or desks.

Security measures, obviously, will vary with the size and nature of your department or agency. Your responsibility is to be sure you exercise reasonable and prudent care appropriate for your own situation. Tight control is a nuisance, but it is a lot less of a nuisance than conducting an investigation to determine how a test or answer key was lost, or how a student's paper was altered while it was in the hands of the training division.

Practical Testing

Student skills must be measured with practical tests. Here, as always, we test to objectives. If an objective states: "Given an operational breathing apparatus, wearing full turnout gear except for gloves and helmet, the firefighter will don the breathing apparatus and be breathing from the air tank, performing all steps completely and correctly, in not more than thirty seconds," the only *effective* way to test ability to meet that objective is by measuring the stated performance under the stated conditions. Practical testing is used primarily in evaluating basic skills, but it can, and should, be used at any point on the career ladder where a student demonstration of personal skill provides the most effective tool for measuring ability to perform.

Scoring can be a simple pass or fail, or it can be based on an evaluation of critical main steps and key points in each evolution. Main step and key point evaluation requires that we establish a specific checklist, identifying how and when points will be awarded, and how and when points will be lost. Does an overrun on time, for example, result in failure or only in the loss of a point or points? Such a checklist is useful, even in an extremely small department or agency, to ensure that each student is judged in exactly the same way. In a large operation, with many instructors, a checklist is absolutely essential, for precisely the same reason. Instructions given to the students must be clear and unmistakable, and these must be exactly the same for each student tested. The best policy is to write out the exact words to be used in implementing each element of the practical exam and to require the instructors administering the test to *read* the instructions. This requirement drives instructors absolutely wild, but it is uniform, fair, and legally defensible. If the evolution being tested has several parts or is lengthy, divide the instructions and issue them in appropriate steps or stages.

It is generally considered best not to discuss *scores*, as such, with individual students during the exam. We do agree, however, with the thought that each student should be privately advised of his or her errors and the corresponding correct practices before being released from the test area. The rationale behind this suggestion is that the test should be used as a teaching opportunity to hammer home correct techniques at a time when the student has a vital interest in what is being said. The law of intensity, remember?

It may seem too obvious to mention, but keep students who have not yet been tested in a location where they cannot see or hear what is going on in the test area. In addition, keep students who have completed the practical test separated from those who are still waiting for their turn. Pure common sense, you say? Certainly it is, but we have seen both situations ignored or handled so poorly that they might as well have been ignored. Just file the ideas away, and don't become careless.

Practical testing takes a lot of time. Set up a sufficient number of testing "stations" with enough individual evaluators to keep the total time of the exam session within reason. Give some serious thought to how students will be staged into and out of the testing process. If you are testing twenty-five students in practical skills, and you waste two minutes on each transition, that's fifty minutes lost unnecessarily. What's that? It's only forty-eight minutes? Well, yeah, you're right; we just wanted to see if you were paying attention.

Student safety is absolutely essential. In setting up any new practical exam, it is not a bad idea to walk through the entire test once or twice, using instructors or other experienced personnel in place of students, to make sure there are no "booby traps" which might injure a participant or damage any of the gear being used.

Student Evaluation Reports

The formal evaluation process often includes subjective instructor judgment of student characteristics which cannot readily be measured with a written or practical test. Most often, instructors are asked to evaluate such items as cooperation, willingness to participate, willingness to work, observation of safety rules, and good old, elusive, "ability to get along." We feel that, under certain circumstances, these judgments may be quite important, but you may find yourself and your agency on shaky ground unless highly specific evaluation criteria, and clear rules for their application, are documented in writing first and are explained in advance to all students. Even then, such evaluations should probably be officially requested and recorded only during entry-level courses and probationary periods when discretionary instructor evaluation is ordinarily expected and generally tolerated.

The Privacy Act and Grades

For generations it has been common practice to post a grade list on the classroom door as a means of advising students of their grades in exams or courses (Fig. 9-7). The list, of course, told each student not only what grade *he* or *she* had scored but what everybody *else* had, as well. Today, this practice is no longer legal. The grades of a student must be treated as private information, and they may not be disclosed or transmitted to a third party without the express consent of the student involved. A class registration card on which a student firefighter signs a release, permitting the instructional agency to provide grade information to the student's fire department, has been deemed invalid by legal counsel. We recommend that you mail grades to the student's home address, and that you not disclose or release grades or pass-fail information to any third party without specific written instructions from the student.

Test Analysis

Probably the most common method of analyzing student examination scores is to calculate the arithmetic mean or "average" grade for the class. The total number of points scored, divided by the number of students taking

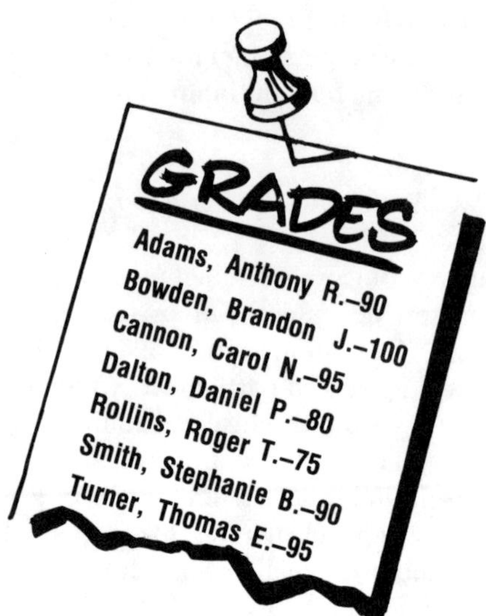

Fig. 9-7. Public posting of grades by name is not legal.

the exam, equals the average score. Within that particular class, the average score provides a base for assessing the relative proficiency of a particular student as compared with the peer group "average" for that class. From class to class, the average score gives us an idea of the relative performance of different groups and, to a degree, the relative performance of different instructors.

Assuming reasonably homogeneous groups of students, an instructor whose classes consistently score an average of five or ten points above the average of all similar classes is clearly doing *something* better. Conversely, an instructor whose classes consistently score an average of five or ten points below the average of all similar classes is clearly doing *something* poorly. In either set of circumstances, we should try to determine what that *something* is.

A sudden major drop in the average score of a single class taught by a usually reliable instructor is cause for curiosity. A major drop in the average scores of several consecutive classes taught by that same instructor is cause for concern.

If your review of class averages seems to suggest that there is a problem with a particular instructor, make haste slowly. Follow your instincts, by all means, but beware of making quick judgments or rash statements. A quiet look at the situation may be all that is needed; what appears to be a statistical mountain in your office may turn out to be only a molehill in the classroom. Don't go off half-cocked.

You must be careful of reading too much into averages, as you must be careful in listening to statisticians. If you stand with one hand in a bucket of boiling water and the other hand in a bucket of ice water, a statistician would probably argue that, "on the average," you should be quite comfortable.

Graphs or Histograms

Another common way to gain information from exam scores is to plot them on a graph, or histogram, on which the vertical axis represents the number of students achieving a particular score and the horizontal axis, the grade point score from zero to one hundred (Fig. 9-8). In conventional educational settings, the analyst would expect to see a classic bell-shaped curve, with most students piling up somewhere between 70 percent and 80 percent, trailing down from that peak, on both sides, in a regular, diminishing progression to a few very low and a few very high scores. Unfortunately, some of our colleagues in fire service education and training think this is how *our* student scores should occur on a plot. We disagree.

Back in Chapter 1 we talked about the differences between our responsibilities and those of other educators. We said in slightly different words that we, as fire service instructors, had to see, to the best of our abilities, that our students learn, to the best of their abilities. We believe that statement clearly indicates we should *not* be anticipating a bell-shaped curve; instead, we should be striving for a skewed curve on which our students pile up at the high end of the scale. Perhaps a practical example would help.

95 Percent of the People at 95 Percent Proficiency

Many years ago we read, in a DuPont company safety publication, an account of their corporate efforts to reach all of their supervisory personnel with a "Supervisor's Safety Course." When the course was first introduced and the scores of the initial supervisory groups were analyzed, corporate safety officials were disappointed to see bell-shaped-curve plots (Fig. 9-9). The material was far too important to settle for average levels of proficiency; they *did not want* "bell-shaped-curve" results. The goal of top management was that 95 percent of their supervisors should score 95 percent or better. Their training and safety personnel, working together, kept reworking, rewriting, and retesting the program until they got the results they were after: 95 percent of the students mastering 95 percent of the material.

Let us be quick to point out, DuPont did *not* force high performance grades by watering down the examination process—nor should *you*. High scores are not the issue. The issue is *learning*. You push and pull student

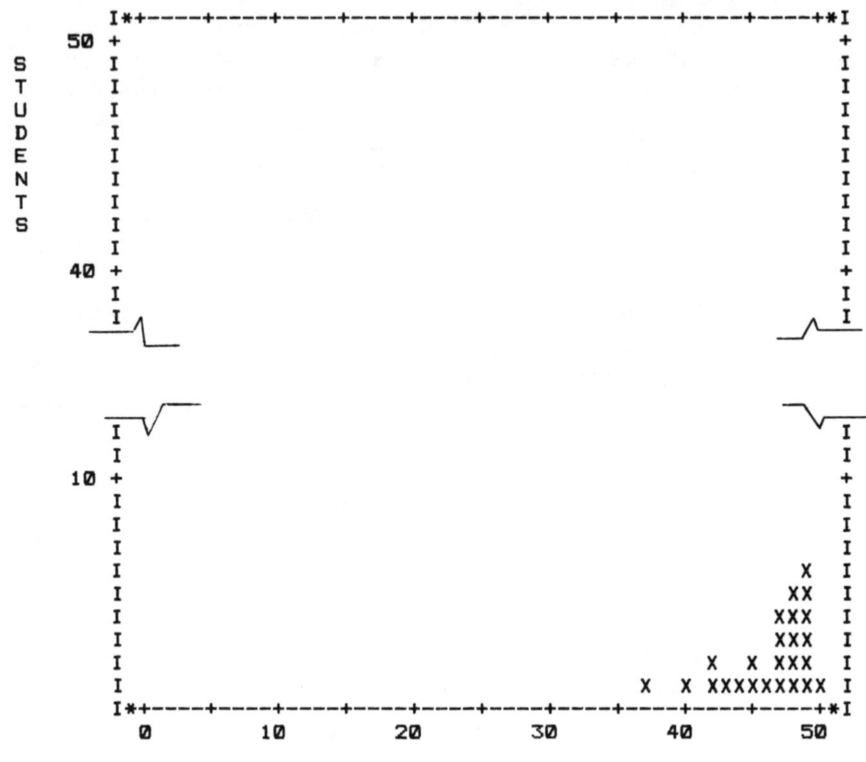

Fig. 9-8. Histogram.

scores and *student learning* to the high end of the scale by evaluating and rewriting performance objectives, by extending classroom instructional time, by manipulating instructional methods and instructional media, and by motivating.

Reading that article, we thought, "Wow! That's what the fire service should be shooting for." We've never changed our minds. We fully realize that, in an entry-level program with no preliminary screening, such a high goal is not realistic. Nevertheless, the tighter your entry screening process, and the farther students advance through your training program, the higher your standards should become. You know, and we know, those standards are not always as high as they should be or could be.

In all honesty, in our opinion, the fire service has been willing and able

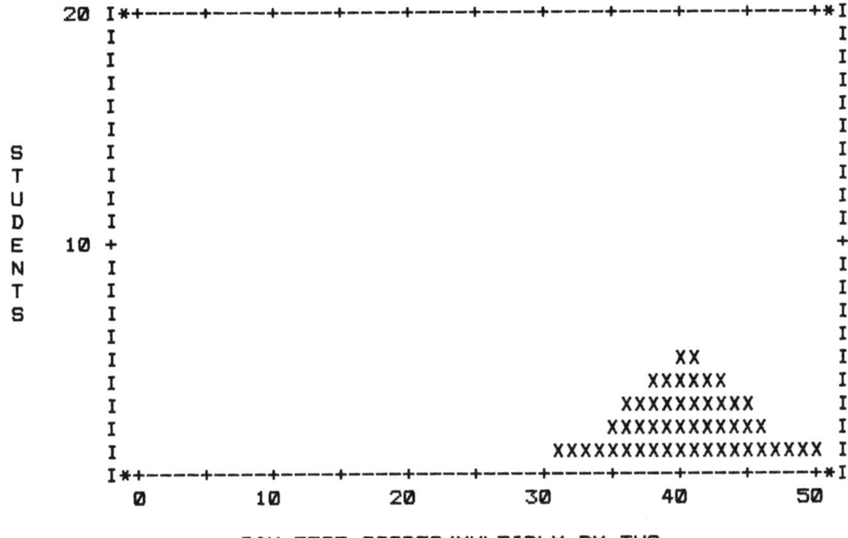

Fig. 9-9. Bell-shaped curve.

to carry along a lot of "dead wood" because our operations are essentially team efforts in which the better participants pick up any slack caused by the less effective members. Observing some of the more positive and progressive aspects of emergency medical service training (EMS) has caused us to take a fresh look at standards for individual student proficiency. The EMS person in the back of an ambulance is in a "one-to-one" relationship with the patient under his or her care. That EMS person *must* have the ability to perform *all* essential skills, to perform them quickly, precisely, and effectively. Testing of EMS student knowledge and practical skill tends, therefore, to be measured by standards which are much higher and which bend or skew their performance curves to the high side, at least in critical areas of practical skills (Fig. 9-10). Not everybody passes, of course, but those who do complete can perform essential skills at the high end of the scale.

We feel there is an important lesson to be learned from our brothers and sisters in emergency medicine and from the DuPont company. Between you and your top management team you can establish whatever standards you decide are appropriate: 70 percent of your students at 70 percent proficiency or 95 percent of your students at 95 percent proficiency or any point in between. What are *you* willing to settle for?

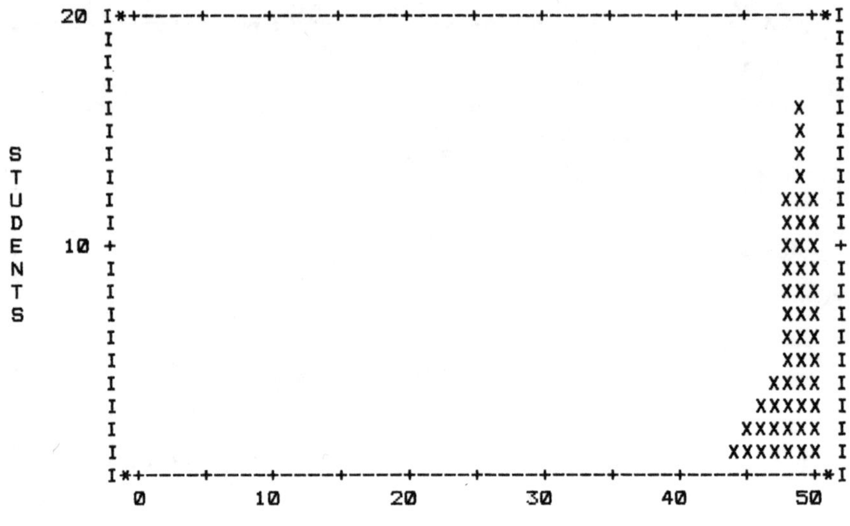

Fig. 9-10. Curve skewed on the high side.

Analyzing Questions and Tests

If we are correct in believing that the standard bell-shaped curve is *not* what we are looking for, we cut ourselves off from quite a few of the classic statistical tools normally used by educators. For example, "standard deviation," a way of measuring how well an examination is bringing an appropriate number of student scores into the center of the bell-shaped curve, becomes relatively meaningless to us. There are, however, a couple of simple analytical studies we can profitably make.

Item Difficulty

One useful study is to analyze item, or question, difficulty. For each class, a simple clerical or machine count is made of the number of students, or the percentage of students, correctly answering each question. These class counts are aggregated to give us, for each question, the number or percentage of correct answers for all classes. We can then compare questions within a single class and compare the results within one class to the norm for all classes (Fig. 9-11).

```
                    I T E M   S T A T I S T I C S

-------------------------------------------------------------------------------
ITEM DIFFICULTY      (PROPORTION OF RESPONDENTS ANSWERING CORRECTLY)

   1  -   5    1.00         .90        1.00        1.00        1.00
   6  -  10     .65         .90         .75        1.00         .95
  11  -  15     .90        1.00         .90         .95         .90
  16  -  20     .95         .95         .90         .80        1.00
  21  -  25     .95        1.00        1.00        1.00        1.00
  26  -  30     .90        1.00         .85         .85        1.00
  31  -  35     .80         .85         .85         .95         .80
  36  -  40    1.00        1.00        1.00        1.00        1.00
  41  -  45     .95         .75         .35         .90         .95
  46  -  50    1.00        1.00         .80        1.00         .95


-------------------------------------------------------------------------------

ITEM OMISSIONS

   1  -   5      0           0           0           0           0
   6  -  10      0           0           0           0           0
  11  -  15      0           0           0           0           0
  16  -  20      0           0           0           0           0
  21  -  25      0           0           0           0           0
  26  -  30      0           0           0           0           0
  31  -  35      0           0           0           0           0
  35  -  40      0           0           0           0           0
  41  -  45      0           0           0           0           0
  46  -  50      0           0           0           0           0


-------------------------------------------------------------------------------

ILLEGAL RESPONSES

   1  -   5      0           0           0           0           0
   6  -  10      0           0           0           0           0
  11  -  15      0           0           0           0           0
  16  -  20      0           0           0           0           0
  21  -  25      0           0           0           0           0
  26  -  30      0           0           0           0           0
  31  -  35      0           0           0           0           0
  36  -  40      0           0           0           0           0
  41  -  45      0           0           0           0           0
  46  -  50      0           0           0           0           0


-------------------------------------------------------------------------------

           TYPICAL COMPUTER ITEM ANALYSIS - FIFTY-ITEM TEST

The low success rate on question 43 may indicate only that this particular
instructor failed to adequately cover that point with this student group
or, if the low percentage is noted in number of classes taught by differ-
ent instructors, it may indicate a poorly constructed question.
```

Fig. 9-11. Item analysis.

A high percentage of correct answers, something approaching 100 percent, might indicate that a particular question is too easy, but it might
indicate instead that we are doing our job very well. A low percentage of
correct answers may have a variety of meanings. A low percentage of correct
answers across many classes may mean that the question is too difficult, that

the question is poorly written, or that the material which the question addresses is not well defined in the official lesson guide.

A low percentage of correct answers to a particular question in one class, while most classes have high percentages of correct answers to the same question, very probably indicates the instructor of that class did not cover that objective well, or perhaps overlooked it entirely. A high percentage of correct answers to a particular question in one class, while most classes have low percentages of correct answers to the same question, may indicate the instructor of that class has a particularly good technique for teaching that point or, more likely, that the instructor is "teaching to the test."

Teaching to the Test

"Teaching to the test" is the practice of spending extra time or placing extra emphasis on those points which the instructor knows will be on major examinations. Whether or not it is improper depends on the degree to which it is done. There is nothing inherently wrong with emphasizing major points. On the other hand, it *is* wrong to carry that emphasis to a point where it totally compromises the testing procedure. *That* practice must be stopped.

Item Discrimination

The usefulness of a particular question in effectively separating the better students from the poorer students is called "discrimination." The discrimination power of a question can be measured by comparing the number of correct answers scored on that question by students in the top half of the class with the number of correct answers scored by students in the bottom half of the class.

There are a number of ways this discrimination can be calculated. The simplest is to split the test scores for a class in the middle, creating, on the basis of test scores, an upper half, containing the top 50 percent of the students, and a lower half, containing the rest. For each item or question, count the number of correct answers in the upper half of the class and the number of correct answers in the lower half of the class. For each question, subtract the number of correct answers in the lower half of the class from the number of correct answers in the upper half of the class. The result will be a positive number, zero, or a negative number. If you just want to make a quick inspection of that one class, this procedure is all you need do. This raw number represents the discrimination power of any particular question as compared with the other questions within that test and within that class. It is limited, however, by the number of students which make up half of the class and can be compared only with other classes of exactly the same size. Dividing that number by the number of students in the class will convert it to a percentage which can then be compared with other tests and other classes of any size.

A positive number or percentage is an indication that the question does discriminate; the higher the number, the higher the discrimination power. A zero indicates that the question is neutral with respect to discrimination, although it may still be useful for other reasons. A negative number indicates that something is wrong. If more poor students than good students are answering a question correctly, there is either something definitely misleading in the question or, for some reason, everyone is guessing. If the same negative discrimination results show up for that question across a number of classes, it should be rewritten or eliminated.

Validity

Let's return for just a moment to a term we introduced at the beginning of the chapter: question and test validity. The validity of a question is tied tightly to the validity of the objective which it is measuring. If the objective is correctly structured and clearly written, there is a high probability that a properly written test question based on that objective will be valid, that it will measure what it is supposed to measure.

A valid test, based on valid objectives, should correctly predict student occupational success. The validity of the examination process, therefore, can be measured against operations in the field (Fig. 9-12).

Suppose the training division conducts a departmental course in the basics of sprinkler system operation and the test results indicate 95 percent of the students are 95 percent proficient. If we get out on the fireground and suddenly discover that none of our people know how to put a simple wet pipe sprinkler system back in service after a fire, the validity of *that* final examination, or the course itself, suddenly becomes extremely suspect.

Inspect the final test scores and see how closely student performance on the examination agrees with what you expected. We said earlier that most instructors have a pretty good idea, long before the final, of how they expect individual students to place. If there is a wide discrepancy between what you expected and what the test indicates, we would be willing to bet that *you* are right and that the validity of the test is questionable.

Look at the score of each individual student and compare it with scores the same student received on other examinations in comparable subject areas. Similar scores indicate a high probability that the current test you are

Fig. 9-12. Does it measure what it is supposed to measure?

reviewing is valid. Neither of these methods would be considered "elegant" by professional statisticians, but they are useful "rule-of-thumb" indicators.

How Deep Do You Dig?

An honest answer to the above question is that we don't really know. The development and analysis of tests, like so many other elements in teaching, is more of an art than a science. No matter how you try to quantify any analytical study of examination questions and student responses, no matter how many numbers and mathematical computations you apply, you will, sooner or later, come to a decision point where your personal judgment is all that you can really rely on. If the testing package you are using appears to be delivering acceptable results and it is not raising any significant number of student complaints, there is no real reason to question what you are doing (Fig. 9-13). As the saying goes, "If it ain't broke, don't fix it!"

If you have a genuine testing problem which is causing you or your students concern, your own intuition may be as useful as anything else, and your intuition plus some of the simple analytical tools we have just described may well be all you need to solve the problem.

If you enjoy playing around with numbers and you want to dig deeper, there are, in most libraries of any size, many books on education which contain all kinds of statistical measurement tools for analyzing questions and tests. You will find that most, if not all of them, as we said earlier, will assume the bell-shaped curve to be the ideal distribution of test scores. If you agree with our idea that we are looking for a curve skewed to the high side, you must apply standard educational analysis techniques with care; they may well lead you in the wrong direction.

Instructor Evaluation by Students

One final evaluation item we should mention are forms on which students are asked to evaluate *instructor* performance. Instructor preparation, presentation, motivation, courtesy, interest, discussion leadership, use of training aids, consistency of teaching with stated objectives, and any other instructor qualities are fair and reasonable points on which we can solicit student opinion.

Fig. 9-13. How deep is deep enough?

If instructor evaluation forms are handed by the student evaluator directly to the evaluated instructor, usually along with the student's final examination paper, they are virtually worthless. Students placed in this situation are simply not going to attack the qualities of the person holding the class grade book. (We don't think *you* would either.)

For student evaluation to have any real merit and honesty, there must be a reasonable guarantee that any opinion expressed will remain anonymous. This fact makes it imperative that any system you establish must bypass the class instructor. One method is to provide the necessary forms and a large envelope in which they can be sealed. Completed forms can be collected by a student, placed in the envelope, sealed, and then turned in to the instructor, or, better still, turned in directly to the training office.

We have found student evaluations to be of help in assessing our teaching and in evaluating the merit and structure of courses and seminars. Most fire service people realize that they have a vested interest in improving the quality of instruction and of course content, and they are usually willing to take a few minutes to jot down their thoughts. If you give them a proper and convenient method for doing so, they will "tell it like it is."

Well, it's getting late in the day and we want to get you out of here and on your way home before the evening traffic builds up. There's just one thing more we think we ought to mention. We referred to computers a couple of times in different contexts. We talked about word processing in course and lesson development, interactive computer programs as one method for generating student application, and, in this chapter, the advantages of computer-driven grading and test analysis. Before you leave, let's stop back by the office and we'll take a quick look at the world of "hardware" and "software" and tomorrow.

10

Computers

Let's set the record straight at the beginning. We were *not* among those who welcomed computers. We were pulled into the computer age by the technology explosion, dragging our heels and explaining to anyone who would listen why we didn't need to change (Fig. 10-1). Happily, no one paid any attention. We were wrong.

We grew up in this business in an era when state and county training records were hand-posted on three-by-five cards, sorted by fire department, and alphabetized. A telephone call from one of our students, requesting information from that training record, could be answered immediately by pulling the appropriate card. All we could see when computers were discussed initially was the loss of that immediacy. It was hard to give up a system which still worked and which had served us well.

Today, just a few years later, the idea of manually posting the records of our students brings to mind a picture out of Charles Dickens—stand-up desks and quill pens. As we said back in Chapter 1, "Old dogs can learn new tricks." The writing of this book began with a yellow legal pad and a handful of number 2 pencils. By the time the initial chapter was finished we thought, "This is idiotic!" We took a break from writing, did a little market research among fire service friends, bought a PC and a printer (Fig. 10-2), and taught ourselves word processing—a decision we have had absolutely no reason to regret.

Fig. 10-1. Personal computer components. *Courtesy International Business Machines Corporation.*

The Possibilities Are Limitless

We are, therefore, in no way "computer experts." We have no special knowledge or abilities on which we can draw to write a technical chapter on computers or computer programming. Instead, let us then discuss areas where we can see clear and profitable applications of computer technology to training and education.

There is no question that computers currently form the most dynamic part of the entire fire service training picture. The technology changes daily. What was impossible yesterday will be commonplace tomorrow. We are, ultimately, restricted only by our imaginations, our budgets, and the available time—the possibilities, truly, are limitless.

Word Processing

Word processing is one of the simplest and more advantageous computer applications for any educational agency. It is, in fact, the single most

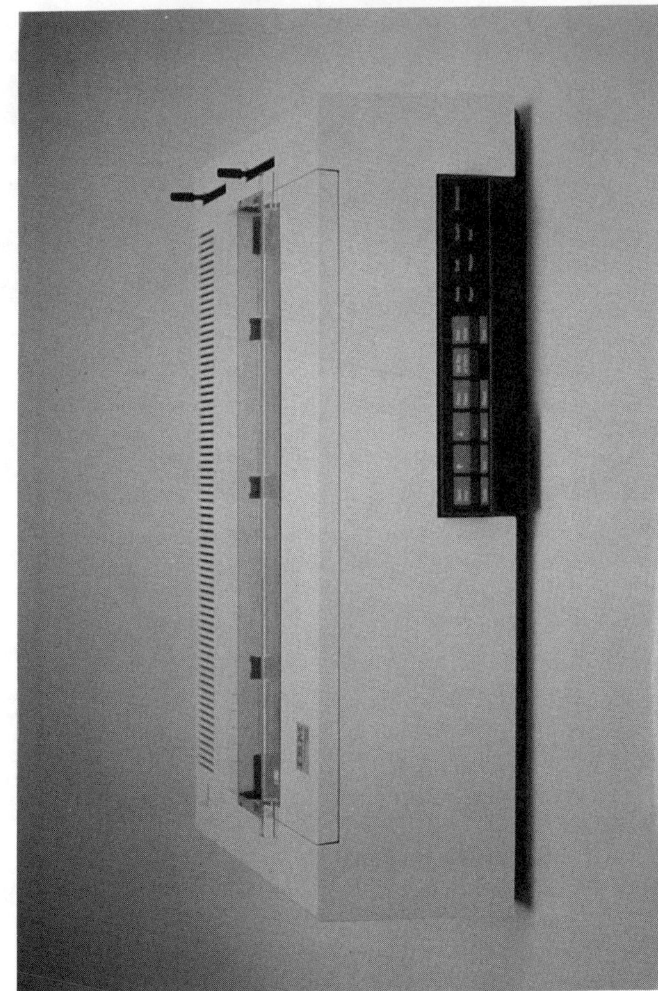

Fig. 10-2. Personal computer printer. *Courtesy International Business Machines Corporation.*

254

common function and use of microcomputers. The ability to manipulate the written word—to insert words, to delete words, to move words—gives us a freedom in creative writing which must be *used* to be appreciated. A word processor, however, can only record what you put in. It cannot, and will not, make up for an inability to write well. On the other hand, we once read that the use of word processing leads to inferior or sloppy writing, just because it *is* so easy to make changes. We seriously doubt that idea. We find that word processors provide a mechanism to get thoughts down rapidly and to polish the resulting words until they say precisely what was intended to be said—and we have written enough pencil and typewriter copy over the years to state flatly that, in the past, a great many poorly written words were shoved out the door just because it *was* so difficult and so laborious to *change* them, once they were inscribed on paper.

We mentioned word processing under lesson planning and course development and said for those charged with the responsibility for course development that it is not a luxury, it is a necessity. If there is anyone left out there in fire service education and training who is *not* using word processing, we hope this book will change your operation. The tools are simply too valuable to ignore.

There are "outline" programs available which lend themselves to blocking out initial thoughts for courses or text materials. These programs include "built-in" standard outline formats and are expandable, accepting additional thoughts and ideas as these are hammered out. Some lesson planners like to use them in developing course outlines; others simply use the keyboard and screen as "pencil and pad," working in some standard word-processing system format.

When the transition is made from outline to lessons, we urge that all work, from as early as possible, be done in the finished page format (Fig. 10-3). It is just as easy to lay out your words from the beginning in what will eventually be the published arrangement, and this method saves a tremendous amount of work later on. With the wonderful advantage of "automatic page numbering," and the ability to insert uniform "headers" and "footers" (repetitive wording and/or lines at the top and bottom of each page) it is a simple matter to set up a standard format for either instructor's guides or student workbooks. It makes sense to us to work in that same page format from the beginning, rather than to cut and fit in two stages. Your operation may be different—if it works, don't change it.

Obviously, student handout material, course announcements, journal articles, letters—any and all word-oriented output—can be generated in the same manner and with the same ease. We have become so "spoiled" that we use the computer for virtually everything we write.

Fig. 10-3. Word processing—the most common function. *Courtesy International Business Machines Corporation.*

Which Word-Processing System?

There are a number of excellent word-processing systems available. We are not going to attempt to tell you which one to buy—but we will suggest a number of points to consider. First, there is no such thing as a magic "all-purpose" system. Each of the major, recognized programs has some relative strengths and some relative weaknesses. System "A" may do one or two things better than system "B," and system "B" may do one or two things better than system "A." For example, some systems adapt to column arrangements more easily; some are more highly "page-oriented"; some are "menu-driven" and tell you what your options are at each change point while others do not. The "Catch-22" situation we all face when selecting a word-processing system is that *before* you start using one, you don't know what questions to ask, and *after* you begin using a particular system, it is not always easy to shift gears and "re-learn" a new one (Fig. 10-4).

If you are planning to purchase word-processing capability, we urge you to talk with some knowledgeable, local fire service people who are doing

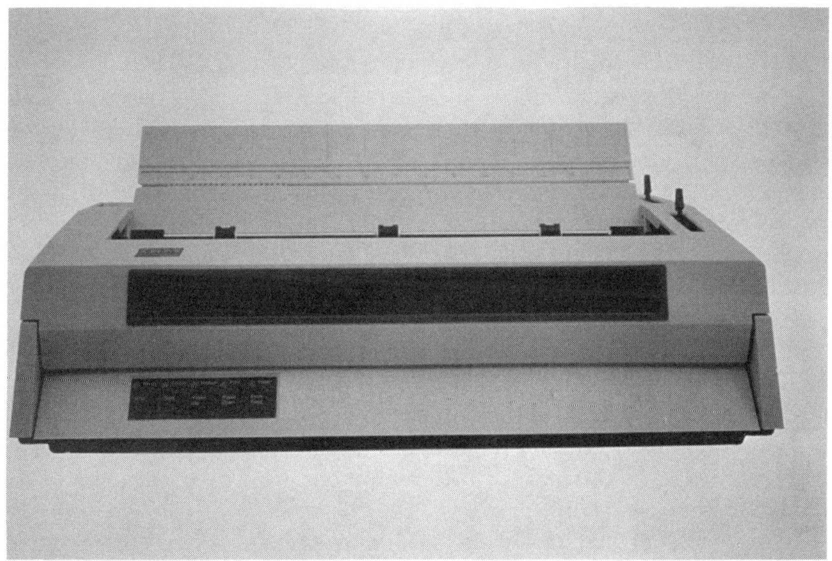

Fig. 10-4. Choose your hardware and programs carefully. *Courtesy International Business Machines Corporation.*

work similar to yours. Think carefully about their suggestions and, if possible, observe the use of their systems in operation—both in creating and laying out material, and in printing it out. Flexibility in laying out pages and speed of operation are both critical areas to consider. The system you finally choose need not be infinitely flexible—none of them are—so long as it has the flexibility to accomplish your purposes. "Speed of operation" would be difficult for us to define. What we have in mind is that most computer packages, hardware and software combined, seem incredibly fast when we first begin to work with them, but they seem slower and slower as our skills increase and the workload backs up. If you have the option in selecting your system, in both programmatic transitions and in printing, quicker is better.

Keep an open mind, to the best extent possible; don't become too dogmatic too quickly. An investment in computer hardware and in computer programs is not only a major expenditure financially, it ties you to a specific system for a long time to come. Make haste slowly. Also, be ready and willing to learn system "B," if you find that system "A" just won't perform some of the procedures you need to accomplish in your operation. If you have the time and patience, there is a tremendous advantage in learning, to a high level of competence, more than one word-processing system. You can then select and utilize the system with the features necessary for the current project. Be flexible.

Storage

Storage of course materials on disks provides a twofold advantage. First is saving space. One recently completed course occupies five three-ring binders on the bookshelves at the office. One binder contains notes on original source materials, correspondence, and other similar hard-copy, background information. The other four contain the originals of the instructor's guide, student workbook, student handout materials, and transparency masters, all of which are also stored on three 5¼ inch floppy disks. The information contained would probably fit on *one* disk, but it was just more convenient to leave them as they were developed.

The *primary* advantage comes when course revision or course updating is required (Fig. 10-5). At this point, our ability to return to the original disks to revise, delete, insert, and move words saves immeasurable time, money, and aggravation. Nothing in the "pre-computer" world gave us anything like the flexibility we now enjoy.

One word of caution—disks can be erased, destroyed, or rendered otherwise useless in an amazing variety of ways. Many years ago, in an aircraft plant, we saw a poster which read, "The sky, like the sea, is terribly unforgiving of our mistakes." Perhaps an appropriate paraphrase would be,

Fig. 10-5. Advantages: space-saving and course updating.

"Computers, like the sky and the sea, are terribly unforgiving of our mistakes." Be sure you have back-ups for all your materials on other floppy or hard disks, and don't store the back-ups in the same place that you store the originals.

Should something important be accidentally erased, stop immediately. Based on our experience, Murphy could write a law stating, "The probability for accidental erasure of a computer disk increases proportionally with the importance of the material stored on the disk." Do not discard or reuse the disk until you have obtained some expert technical advice on the problem. Under some circumstances, it *is* possible to retrieve part or all of the information which appears to be gone but is merely "hidden" from untrained eyes like ours and possibly yours. We certainly can't tell you *how* to do it, but it is quite likely that it can be done.

Our resident computer expert suggested a corollary to our proposed addition to Murphy's law: "The more important the material, the more you fiddle with it; therefore, the more chances for a *Whoops!*" You are now duly warned.

Graphics

Fire service people have for some time been using tremendous ingenuity to blend computer-generated text and graphics into visual aids (Fig.

Fig. 10-6. Graphics continue to improve with each new generation of equipment.
Courtesy International Business Machines Corporation.

10-6). The possibilities, again, seem almost limitless. Color combinations, type font, and type size are all variable, allowing extremely wide latitude in creating title blocks and other art work of true excellence.

There are significant differences in the adaptability of computer hardware to the creation of graphics. Some units are extremely strong in graphics; others are not. Again, your best bet is personally to examine the range of equipment available and talk with your peers who are already involved. Some hardware systems and accompanying programs permit drawing directly on the screen with an electronic stylus, commonly known as a "mouse." Other units employ "desk-top" drawing on a "digitizing pad" which captures the lines created and enters them into the computer storage system. Technically, we are not too far away from direct scanning of maps or drawings, transferring material directly into the computer data base.

Class Administration and Student Records

Every fire service training and education agency has both a legal and a moral responsibility to maintain records on the participation and performance of each student who moves through the system. With the increasing interest in certification and standards, we have a much greater need to meet this responsibility with a high level of accuracy and with reasonable accessibility for the students we serve.

How Can We Store Student Records?

Student records must begin with identification of the individual (Fig. 10-7). By name? Well, maybe. As one of our friends put it, "Names are nice, but they create problems." He went on to tell us of fire departments with three and four generations of firefighters from the same family in the student record books—many with the same or similar first name and middle initials. Social security numbers probably provide us with what we need: a designation, guaranteed unique, for each of the people we teach. For state, county, and regional training systems, then, we need to capture and record the name, social security number, and departmental affiliation as basic information for each individual

Course Identification

For generations, fire service training programs have been relatively casual in recording any specifics about the courses completed by participants. "Basic Firemanship," or some variation on those words, for example, has been used for years in some agencies to indicate the fact that an individual has completed some level of fundamental training. Standards and certification again force us to answer some tough questions. What *was* "Basic Firemanship" in 1952? Or in 1972, or 1982? What subjects were covered? How many clock hours were spent on each? What standards of performance were required?

We need, then, beginning with the offering of each new course or major course revision, to record enough information about that course to be able to state accurately, five, ten, or twenty years from now, exactly what was

```
             CENTER COUNTY FIRE AND RESCUE ACADEMY

                   TRAINING ATTENDANCE REPORT

                   BASIC FIREFIGHTING COURSE


Unit or Course Number:   BF-3-88    Class Location:   NORTH SIDE STATION

Session Number: __________   Date: __________   Instructor:  JOHN J. JOHNSON

Subject covered this session ______________________________________________

------------------------------------------------------------------------------
        STUDENT NAME       :  SIGN-IN  :          STUDENT SIGNATURE
                           :   TIME    :
------------------------------------------------------------------------------
  1   ADAMS, ANTHONY R.     :          :
------------------------------------------------------------------------------
  2   BOWDEN, BRANDON J.     :          :
------------------------------------------------------------------------------
  3   CANNON, CAROL N.       :          :
------------------------------------------------------------------------------
  4   DALTON, DANIEL P.      :          :
------------------------------------------------------------------------------

 ___________________________________________/\/\___________________________

------------------------------------------------------------------------------
 23   ROLLINS, ROGER T.      :          :
------------------------------------------------------------------------------
 24   SMITH, STEPHANIE B.    :          :
------------------------------------------------------------------------------
 25   TURNER, THOMAS E.      :          :
------------------------------------------------------------------------------

        Total students attending this class session __________

        I hereby certify this roster is correct as submitted.


                          ________________________________
                               Instructor's Signature
```

Fig. 10-7. Computer-generated attendance report.

covered under that course title, at that time. Course objectives, time spent on those objectives, course outlines, references—all should be filed for future retrieval under the appropriate title and date. If the course is necessary for certification, the record should also indicate any limitation or expiration date which applies to that training. Prerequisites, or any specific sequence of classes of which this course was a part, should also be noted.

It is also an excellent idea to keep detailed records for each course offering: the date the course began, the date it ended, who was the instructor, where it was held. All of this information gives some real meaning to the record that John J. Jones, SSN 111-22-3344, West Suburbia Fire Department, completed "Basic Firemanship" in 1982 with a grade of 87 percent.

"Hard-Copy" Files

This stage may be a good place to mention that not *everything* needs to go, or should go, into the computer. Computer records should be established to provide *timely, accessible, summary* information. That John J. Jones, SSN 111-22-3344, West Suburbia Fire Department, completed Basic Firemanship in 1982 with a grade of 87 percent is timely, summary information which should be readily accessible—good information for the computerized record (Fig. 10-8). Most of the other information *about* that course may well be better stored on a couple of pages of paper in an appropriate file or, perhaps, eventually, in a departmental microfilm or microfiche system.

There is definitely a place for "hard copy." Putting information into computer records and retrieving information from computer records is costly. Be sure your computer record-keeping system is not only operationally effective but *cost-effective* as well.

Registration Sequence

The initial establishment of a computer record for a given class is commonly called the "registration sequence." At the opening session of each class, students usually fill out an appropriate registration card or form containing spaces to record all the information the offering agency needs and wants. Back at the central office, the information from these cards, entered into a simple computer-driven registration program, will provide the necessary class administration forms: a class roster, a session attendance form, and a company or fire department breakdown.

Gradebook Sequence

Following registration, a copy of the attendance form, filled in by the class instructor and returned after each session, becomes the basis for entering into the computer "gradebook sequence" student presence or absence for each class session—usually a matter of bringing up the class record, entering the session date if necessary, and entering a P (present) or an A (absent) for each individual (Fig. 10-9). This record, of course, establishes that any given individual did, in fact, meet the minimum attendance requirements.

The gradebook sequence also (rather obviously) includes entering grades. Scores on quizzes, projects, mid-course examinations, and final examinations may each be entered separately, or only a final, aggregate score may be recorded. A baseline gradebook program should provide a final

```
CENTER COUNTY FIRE AND RESCUE ACADEMY

BASIC FIREFIGHTING COURSE
Unit or Course Number:    BF-3-88
Instructor:    JOHN J. JOHNSON
Class Location:    NORTH SIDE STATION


CENTER CITY FIRE DEPARTMENT

CANNON, CAROL N.                          211-12-1212
ROLLINS, ROGER T.                         212-23-2323
SMITH, STEPHANIE B.                       213-34-3434


EAST SIDE FIRE DEPARTMENT

ADAMS, ANTHONY R.                         214-45-4545


NORTH SIDE FIRE DEPARTMENT

BOWDEN, BRANDON, J.                       215-56-5656
DALTON, DANIEL P.                         216-67-6767
TURNER, THOMAS E.                         217-78-7878
```

Fig. 10-8. Computer-generated department breakdown.

printout showing, for each student, pass-fail and grade, plus the average grade or percentage score for the entire class. We should probably note that the computer-based "gradebook" is kept in addition to the traditional hard-copy records which instructors maintain for their own information.

How Much Is Enough?

Nobody can answer that question but you. The possibilities are, as we have said, virtually limitless. What you do *with* your computer record-keeping system, what you want your computer record-keeping system to do *for* you are entirely a matter of your needs and your capability. Some obvious additional areas to consider are the ability to develop and print out student transcripts, the ability to interface educational records with certification standards, and the monitoring of individual or company participation in mandatory subject areas. Some fire service training agencies have problems

```
Page 1              CENTER COUNTY FIRE AND RESCUE ACADEMY

                 BASIC FIREFIGHTING COURSE CLASS ATTENDANCE ROSTER
Unit or Course Number:    BF-3-88              Instructor:   JOHN J. JOHNSON

------------------------------------------------------------------------------
      SESSION NUMBER        : 1 : 2 : 3 : 4 : 5 : 6 : 7 : 8 : 9 : 10: 11: 12:
------------------------------------------------------------------------------
  1  ADAMS, ANTHONY R.      :   :   :   :   :   :   :   :   :   :   :   :   :
------------------------------------------------------------------------------
  2  BAKER, BRANDON J.      :   :   :   :   :   :   :   :   :   :   :   :   :
------------------------------------------------------------------------------
  3  CANNON, CAROL N.       :   :   :   :   :   :   :   :   :   :   :   :   :
------------------------------------------------------------------------------


Page 2
------------------------------------------------------------------------------
: 13: 14: 15: 16: 17: 18: 19: 20:EVAL:EXAM:EXAM:FINAL:     STUDENT NAME
------------------------------------------------------------------------------
:   :   :   :   :   :   :   :   :    :    :    :     :   ADAMS, ANTHONY R.
------------------------------------------------------------------------------
:   :   :   :   :   :   :   :   :    :    :    :     :   BAKER, BRANDON J.
------------------------------------------------------------------------------
:   :   :   :   :   :   :   :   :    :    :    :     :   CANNON, CAROL N.
------------------------------------------------------------------------------
```

Fig. 10-9. Computer-generated attendance record.

with chronic dropouts: individuals who enroll over and over again in fire service training programs but who seldom, if ever, complete anything. It would be advantageous, certainly, to have the registration system "flag" these people when they sign up, so that they can be dealt with in some appropriate and timely manner.

Ideally, the computer-entered training records within each fire department or training agency should provide the statistical basis for the annual training report. It should be relatively easy to arrange the record-keeping program to furnish, upon demand, subject/hour breakdowns, numbers and/or percentages of students completing specific courses, and company or departmental training achievement records by total hours, total personnel, or both.

Examination Analysis

We earlier mentioned the advantages of electronic scoring of examinations. An electronic scoring system, tied into an analytic computer program, can provide the answers to just about any statistical question you want to ask—providing, as always, that you have the technical and financial resources to do the work and pay the bills.

Some items you may wish to consider might include the analysis of student responses to individual questions in order to provide the number or percentage of students answering each question correctly. In our experience, this is the single analytic element our instructors immediately turned to when the printouts became available. We know of one part-time state instructor who developed a personal program which told him, on the basis of those questions a high percentage of his students missed, what text sections needed additional emphasis or study. A histogram or graphic display of the distribution of student scores is useful, as are some form of analyses of item difficulty and item validity. We've·not seen it done, but it should be possible to provide, on the basis of average class grade, a comparative analysis of an instructor's current performance versus past performance in similar courses, as well as a comparison of the performance of one instructor with others teaching similar programs to similar students.

Security

In Chapter 9 we discussed the need to keep student scores and records confidential. Confidentiality applies to computers exactly as it does to hard-copy files. You must, by whatever means are appropriate and expeditious for your system, ensure that training records—indeed, *all* personnel records— are totally secure and can be accessed only by those with the need and authority to do so.

Computer-Aided Instruction

We feel the most attractive area for fire service computer-aided instruction probably will be found in tactics and strategy gaming and in management gaming (Fig. 10-10). Because of the cost and the resources required, these programs will lend themselves to larger departments and agencies, but, as we have experienced in all phases of computer operations, the programs will be copied, streamlined, and formatted for smaller operational units once they have proven their merit. As we mentioned earlier, the limited interactive computer programs currently available have not appeared to be, at least to us, particularly useful educational tools. Those we have seen do not have a high enough level of sophistication or educational value to

Fig. 10-10. Interactive computer learning. *Courtesy International Business Machines Corporation.*

make them worth the time, trouble, and expense involved. Tomorrow—who knows? And "tomorrow" may arrive sooner than we think. It usually does.

Interactive teaching packages, "self-taught" instruction, or "self-paced" instruction in a wide range of fire service subjects may have a role in the future, but the tremendous cost of development and the limitations of the fire service market make their extensive use doubtful, at least. Again we hesitate to be too dogmatic. The technology has moved so rapidly in the last decade that costs may come down and availability may go up much more rapidly than appears possible from our vantage point today.

One of the standard computer commands is "Escape" and we think it is time to punch the "Escape" key and do just that. As we confessed, our technical knowledge is slim; however, our belief in the utility and in the future of computers in fire service education and training could not be stronger. If you are already a "hacker" of sorts, you are, in all probability, far in front of us in your keyboard ability and your technical expertise. If, however, you have not yet taken the plunge, come on in, the water's fine.

We hate to see this time together come to a close. We've enjoyed talking with you throughout these pages, more than you are probably willing to believe. We truly hope you have profited from the visit, but it's time to let you go. We'll walk with you out to the parking lot and tell you what we would like to see you, as a special and unique fire service instructor, *do* with all the many things we've talked about.

11

The Challenge

We've covered a lot of ground together since you first sat down and we started talking. Let's see; we discussed your role as a fire service instructor, fire service students, the classroom and the drill ground, a little about how people learn, course development and lesson planning, methods of instruction and special techniques for applying what students have learned, training aids, testing, and, finally, an overview of how computers are affecting our field. We *have* covered a lot together, when you stop and look back. We hope some small part of what we've had to say will be useful to you, that it will help you get started if you are just beginning, smooth out some of the rough spots in your existing training program, or simply make your personal teaching efforts more successful and more enjoyable.

You should know, by this time, that we feel *you* are the most important element in your fire service training program. In spite of all we have written about the importance of specific techniques and technology, we believe, with absolute conviction, that a truly caring, creative, motivated instructor can *teach* people, and *reach* people, with nothing more than a hunk of chalk and a bare wall to write on. Top-notch facilities, equipment, and techniques are wonderful things to have, but, in the final analysis, they are only "tools of the trade"—their worth is directly proportional to the skill and dedication of the user. In the hands of a conscientious and skilled craftsman, they are worth their weight in gold. In the hands of an unlearned or uncaring person, they are useless—perhaps worse than useless, because a lot of fancy educational paraphernalia can sometimes create the *illusion* of learning where little or none actually exists.

A book like this one, if you remember Chapter 8, is a single-sense, relatively passive teaching-learning tool. We could, with considerable additional time and expense, take the same information presented in these pages and develop a format for a movie, a video, or an interactive computer program. Any one of these, properly produced, would greatly enhance the effect of our words. But far beyond any medium or machine, the best possible teaching-learning situation we could establish would be to meet each of you face to face. We could, then, together discuss your specific concerns and questions, presenting points, arguing for or against ideas, hammering out solutions, striking sparks from each other, preparing you to meet the problems you will face. *That* is the true role of a teacher in any profession. *That* is the role we sincerely hope *you* will accept as your challenge: to strike sparks from *your* students and to prepare *them* to meet the problems *they* will face. If we have given you anything at all, we hope it is an understanding and an appreciation of the importance of your role in the educational growth and professional development of the people you lead in the classroom and on the drill ground.

We are fortunate, all of us, to be a part of this "gem of all professions." Teaching is a great and rewarding career, and fire service instruction is among the greatest and most rewarding careers in teaching.

One more thing before we close: We sincerely hope you smiled once or twice along the way. If you have worked with us in the past, you know we have been blessed with a rather irrepressible sense of humor. We believe, most earnestly, that students should be encouraged to smile at least once an hour, and that it is not only acceptable but *desirable* that every classroom should ring with laughter from time to time. Don't force it, but don't discourage it, either. (See, there—we *told* you good instructors can't resist teaching right up to the moment their students walk out the door.)

We sincerely hope this book gives each of you as much enjoyment as we have experienced in preparing it for you. We are proud to have played some small part in shaping the fire service of the past and the present. The future is yours. *Vaya con Dios!*

Epilogue

It doesn't seem possible that we have reached the final stages of putting this book together. The work has been a part of my life for several years now, a familiar companion that has dominated my thoughts and time rather ruthlessly. I'll not be sorry to see that come to an end.

Writing this book has provided an unparalleled learning experience. In some respects, I almost wish we could wipe the slate clean and begin right from the beginning—applying the total knowledge gained along the way to the entire project. Unfortunately, that idea would not achieve much success with *Fire Engineering*'s Book Division, who have waited so long for this egg to hatch. I am most grateful to them for their patience, encouragement, and advice and counsel.

Speaking of patience and encouragement, I sincerely thank my wife, Dede, for her constant and dependable support. She shared in my joy when things went well, she held my hand when things went wrong, and she expressed her faith in my ability to complete this work at those times when I doubted I could.

I have been assisted by many people who will not be mentioned by name; I particularly appreciate the generosity of the firms listed in the educational resource section and the courtesy of their representatives, all of whom took time to talk with me and to provide the materials we have included.

One educational resource is worthy of special mention. Through this work, I became acquainted with a publication titled *Fire Service Directory of*

Training and Information Sources, published by Specialized Publication Services, Inc., and available through *Fire Engineering* Book Services. It is a truly comprehensive and most valuable fire service education and training source reference, and it should be in the library of every fire instructional agency.

My thanks to *Fire Engineering,* to everyone who has assisted me in any way, and, in particular, my thanks to you, the reader, for traveling with me this far. It's been fun.

Educational Resources

Charles Mayer Studios, Inc.
168 East Market Street
Akron, Ohio 44308
Telephone: (216) 535-6121

Hook-and-loop boards, hook-and-loop materials, magnetic white writing boards, magnetic materials, wet and dry sprinkler displays, and other fire training aids

Da-Lite Screen Company, Inc.
P.O. Box 137
Warsaw, Indiana 46580
Telephone: (219) 267-8101

Projection screens

Eastman Kodak Company
Rochester, New York 14650
Telephone: 1-800-242-2424

Cameras, film, projection equipment

Elmo Corporation
70 New Hyde Park Road
New Hyde Park, New York 11040
Telephone: (516) 774-3200

Classroom projection equipment

International Business Machines Corporation
2000 Purchase Street
Purchase, New York 10577
Telephone: 1-800-IBM-2468

Computer equipment and systems

Scantron Corporation
1361 Valencia Avenue
Tustin, California 92680
Telephone: (714) 259-8887

Optical grading machines and related forms

Technovate Corporation
910 South West 12th Avenue
Pompano Beach, Florida 33069
Telephone: (305) 946-4470

Fire training simulators